THE COMPLETE
Caving Manual

THE COMPLETE
Caving Manual

Andy Sparrow

THE CROWOOD PRESS

First published in 1997 by
The Crowood Press Ltd
Ramsbury, Marlborough
Wiltshire SN8 2HR

enquiries@crowood.com

www.crowood.com

Revised edition 2009

This impression 2020

British Library Cataloguing-in-Publication Data
A catalogue record for this book is available from the British Library.

ISBN 978 1 84797 146 3

Disclaimer
Please note that the author and the publisher of this book are not responsible in any manner whatsoever for any error or omission, or any loss, damage, injury, adverse outcome, or liability of any kind that may result from the use of any of the instructions/information contained in this publication, or reliance upon it. Novice cavers are advised to join a caving club or seek the guidance of a qualified instructor. Since the physical activities described in this book may be too strenuous in nature for some readers to engage in safely, it is essential that a doctor be consulted before undertaking training for caving, or caving itself.

Typeset by Carolyn Griffiths

Printed and bound in India by Parksons Graphics

Contents

Introduction to the 2009 Edition

More than ten years have passed since the first edition of this book was published and it is interesting to reflect on the changes, discoveries and innovations this decade has witnessed.

The popularity of caving probably peaked in the mid-1970s, generating, at that time, concerns that overcrowding was becoming detrimental to both the fragile cave environment and to the quality of the caving experience. The predominant view of cavers then was that it was undesirable to actively promote caving or encourage over-participation. During the 1980s and 1990s new activities like paragliding and mountain biking provided alternative forms of outdoor adventure and the pressure on caving diminished. We have now reached a point where the number of new cavers, particularly young cavers, has decreased to such an extent that it is causing justifiable concern. Reaction to this new situation is, predictably, divided in the caving community and, while some take the view that 'small is beautiful', it is generally recognized that the institutions of caving, such as the clubs, governing body and rescue teams, cannot be sustained effectively without a new generation of recruits. Thus it is that the British Caving Association now has a youth development officer and actively promotes 'try caving' days.

It is perhaps ironic that while fewer young people are inclined to take up caving as a pastime more of them than ever before are likely to have had a caving experience. As I know through my work as a caving instructor, the majority of young novices find caving highly enjoyable yet very few are likely to repeat the experience. The reluctance of most caving clubs to take responsibility for or mentor young cavers is regrettable but hardly surprising given their concerns over insurance, liability and child protection. By the time a potential caver reaches the age of eighteen the memory of that single caving experience is likely to have diminished as the myriad distractions of our society and its technology hold sway.

Another significant element in the decline of caving is the distorted public perception created by the media. Caving has always struggled, as the 'invisible' sport, with its public image. Back in the 1960s the newspapers delighted in branding us as irresponsible loonies requiring rescue at public expense, which generated considerable hostility amongst their readership. It was the power of television that changed these misconceptions in the early 1970s when excellent primetime documentaries by caving cameraman Sid Perou successfully conveyed

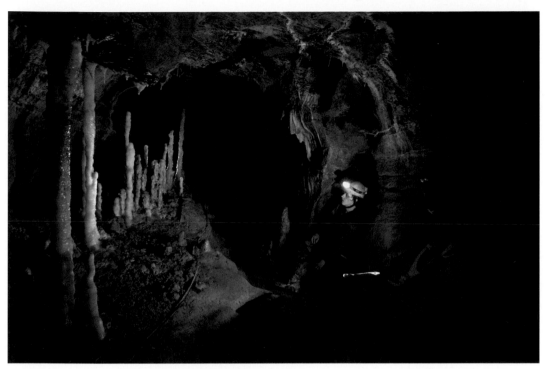

The hidden world beneath the hills. (Photo Rob Eavis)

the reality of what we did and why. Today there are still occasional high quality documentaries about caves and caving but they are likely to be lost amidst the multitude of channels and not to appeal to the DVD and computer game generation.

In this competitive reality it is not surprising that even those who were once the bastions of straight documentary feel compelled to 'spin' and that caving, when it does find airtime, must subscribe to the contrived label of 'extreme sport'. Even worse have been the excesses of the dramatists who find caving useful for storylines in programmes like *Casualty*. It is never enough now to have either a rockfall or a flood in these episodes; both will be required, while any pre-accident caving must be contrived to be both claustrophobic and masochistic. It is a sad reality that the public perception of caving in the

UK is largely inspired by writers who have their own inflexible agenda.

While numbers have declined, we enjoy many benefits that have originated from new technologies. The internet provides an entry portal for today's new cavers, linking them to club websites, discussion groups and cavers' forums. The old caving guide books may be out of print but we can download a survey or rigging topo, or 'Google' for a route description. Caving, an activity that relies so much upon access to information, has never been so well served.

It is now more likely to be a GPS unit than a map and compass that guides us towards an entrance on a misty moor and the caves we once illuminated with a beam of yellow light are brighter but cooler in the bluish aura of our LED lamps. There is no need for that heavy battery with a trailing cable when three or four AA batteries are sufficient for a long day of caving. Those images we strive to capture are less elusive now we can see them instantly displayed on our digital camera and we can always add a touch of Photoshop to make that final enhancement. The cave surveyor no longer needs to squint through the sight of compass and clino when an electronic 'pony' can do the job at the touch of a button; likewise the rustle and winding of the tape measure has been replaced by the bleep of the 'disto'. The drawing board is now the computer screen where we can see and rotate our cave in three dimensions or superimpose it on the latest satellite images.

Another positive and welcome development originates from social rather than technological changes and that is the greater involvement of women. Caving is no longer the male-dominated pastime it once was and it is particularly pleasing to see this reflected in the gender balance of younger recruits. The growing participation of women is a factor that will help caving regain its equilibrium and promote an improved image by harnessing new energies and talents to face both current and future challenges.

During the last ten years many kilometres of cave passages have been revealed under the moors and hills of Britain. The great complex of Ogof Draenen, only recently opened when the first edition went to print, has grown relentlessly and become a world class system over 65km long. In Derbyshire

the magnificent Titan Shaft has been an inspiring addition to the ever-expanding Peak Cavern network. Under the Mendip Hills of Somerset the beautiful and extensive Upper Food Swallet has finally rewarded the diggers who laboured so long and hard to engineer a way through a succession of chokes.

Internationally, spectacular cave discoveries are too numerous to list. Krubera (Voronja) Cave in Georgia has become the first cave to break the two-kilometre barrier, achieving a total depth of 2,191m. Since 1996 ten new systems have exceeded that once 'holy grail' of depth greater than 1,000m. Kentucky's Mammoth Cave at 590km maintains its pride of place as the world's longest, but there are now seven other systems over 160km long.

Here is a final thought. Somewhere deep under the Welsh mountain of Llangatock is an undiscovered link between the two great caves of Agen Allwed and Ogof Daren Cilau. Only a few tens of metres separate these magnificent systems and a linking passage would create not only a 67km system but the longest and most challenging entrance-to-entrance trip in Britain. This link, assumed to be attainable and imminent when the first edition of this book was published, remains tantalizingly elusive. Perhaps the next generation of cavers, inspired by this ultimate prize, will fulfil this vision before the next edition of this book is written.

Why?

Ours is the invisible adventure, for there can be no spectators. The public see us changing by some roadside, stung by rain or sleet, and then, hours later, returning wet, muddy and weary. Their mental images of caving are those created by the TV dramatists, of claustrophobic crawls, falling rocks and surging floodwaters.

Why do we do it? Why do we leave the green hills and blue sky to enter this dark and chill world? For myself the answer is simple enough and can be summarized in four words – beauty, adventure, mystery, comradeship.

The beauty begins even before we leave the surface, for we are likely to be in the midst of a limestone landscape. It may be the gentle green hills of Somerset that surround us or the sweeping vistas of the Yorkshire Dales. We might find ourselves, after a long walk, in the remote Easegill valley, where the pale grey crags are etched and sculpted by the mountain stream or in the heart of the great Welsh wilderness of Mynydd Ddu. If we venture further, to the great caves of France, it is alpine peaks that make a panorama as we traverse the rocky chaos that is the *lapiaz*.

Cave entrances, portals to the underworld, often have their own visual appeal. This may be in the form of architectural symmetry, a perfect tube or gothic arch, or it may be a dramatic chasm of terrifying depth.

Stalactites and stalagmites, in their myriad forms, are much admired by visitors to commercialized caves. While these are often very attractive they seldom match the wealth of colour and complexity that exists in 'wild' caves. Beauty is enhanced by remoteness and challenge. The huge stalagmites of the Gouffre Berger or blue-green helictites of Ogof Daren Cilau can only ever cast their spell over explorers who have made long and difficult journeys.

The halls of the underworld soar into darkness like the naves and domes of midnight cathedrals. They may be tomb silent, or dripping and trickling. Streams and rivers murmur distantly, drawing the caver with a siren's voice towards thunderous waterfalls and rapids.

While 'beauty' is a barely adequate word to describe the ambience and visual appeal of caves, 'adventure' accurately defines the experience of being there. Both the novice caver wriggling through an easy squeeze and the cave diver at the cutting edge of technology will testify to this. Even the most experienced caver will feel the adrenalin surge as they abseil down a deep shaft and such feelings are enormously magnified when the exploration is of a virgin cave.

When the adventure of exploration combines with the mystery of the genuinely unknown the intensity of the experience is unique. It becomes a passion that can take the caver to mountain, rainforest and desert. Such areas offer countless unexplored entrances that could reveal caves of enormous length and depth or exceptional beauty,

Features of caves and limestone. (Rhian Hicks)

but the caver does not need to travel so far to have the possibility of discovery, for even in a country like Britain only a tiny fraction of our underworld has been entered.

Every year in the UK, exploration by digging, climbing and diving expands our knowledge of the underworld. There can be few experiences as satisfying as making the first footsteps in a newly discovered cave. Every twist and bend of the passage reveals some new secret as the explorer's light shines on that which has been forever dark. Excitement mounts with every new turn and chamber, with every junction that offers enticing possibilities.

It is likely to be a team effort that has achieved this goal, often as a result of prolonged labour in harsh conditions. Teamwork and comradeship are essential elements of successful caving. Most cavers have their strengths and weaknesses – some are happy to push a tight squeeze but not to lead an exposed climb, while others are the opposite. It is only by combining our collective abilities and by relying on and trusting each other that we are likely to achieve our objectives.

The dedicated caver has an enquiring mind, a spirit of adventure, and an emotional connection with the natural world. In a world that dilutes and simulates adventure the caver knows the profound satisfaction that overcoming genuine hardships and hazards can generate. And the caver, almost uniquely in our society, can venture into the mystery of the unknown.

Why go caving? Because it will take you to amazing places with amazing people to do amazing things.

CHAPTER 1

The Formation of Caves

Three hundred million years ago, during the Carboniferous period and long before even the dinosaurs evolved, the oceans teemed with life. In warm and shallow seas shellfish and other marine organisms lived, died and decayed amongst the coral. The shells and skeletal remains of these ancient life forms combined on the seabed with other sediments to form a deep deposit rich in calcium carbonate that would eventually become the rock we know as limestone.

Over hundreds of thousands of years the deposits grew and deepened. As climate fluctuated it favoured one species over another and the deposits, with their fossil record, varied accordingly. Sometimes, possibly due to fluctuations in sea level, sedimentation halted and then resumed, leaving a break between the sediment layers. Other events, like a great flood sweeping river sediments into the sea or volcanic eruptions showering ash, added further chemically different layers to the ever-deepening sandwich of deposits. The differing bands, or beds, remained separated from each other by horizontal cracks known as bedding planes.

The huge pressure and stresses caused vertical cracks known as joints to develop. The crystalline structure of limestone causes it to fracture along a natural grid pattern, which is why limestone boulders are usually rectangular and exposed limestone has a brickwork pattern. That joints are formed on this grid pattern can be seen clearly on some surface exposures such as limestone pavements in the Yorkshire Dales, where the smooth, level rock surfaces called clints are incised by deep fissures (joints) known as grikes.

Some masses of limestone became exposed by lowering of the sea level, while others were lifted by isostasy, the gradual rising or sinking of the land mass. In both these cases the limestone remained relatively undisturbed with the original beds still level but other deposits were squeezed under the enormous stresses of plate tectonics (continental drift). These beds were uplifted and contorted to form some of the world's highest mountain ranges. These stresses often resulted in fracture lines called faults that created further cracks and fissures in the limestone.

Limestone is a very common rock, distributed in huge deposits across the globe. It is found in all climatic conditions, from tundra to desert or rainforest, in deposits that may be shallow or kilometres thick. These landscapes vary enormously but have, in almost every case, one common feature – caves.

The Karst Landscape

Few regions illustrate the principal features of limestone landscape better than the Classical Karst of Slovenia, so much so that the word karst has become a general term to describe all geologically similar regions. In

Limestone pavement above Malham, North Yorkshire. The grid pattern of joints is well illustrated. (Andy Morse)

the Classical Karst rivers flow across alluvial floored basins (known as poljes) until they encounter the surrounding limestone hills. Here the rivers enter large horizontal caves, or vanish into sink holes. The rivers flow through extensive cave systems, often several kilometres long, before returning to the surface through a resurgence. There is no surface water on the limestone hills, which are riddled with cave entrances, gaping shafts and funnel-shaped sink holes called dolines. The limestone has little topsoil and is eroded by

OPPOSITE: Ogof Clogwyn: a phreatic passage with rock shelves formed on the horizontal beds of limestone. (Brendan Marris)

water into deep grooves, or karren, creating a complex landscape with much exposed rock.

The Chemistry of Cave Formation

Rain falling through the atmosphere absorbs carbon dioxide to become a weak solution of carbonic acid. The quantity of carbon dioxide is very small (three parts in 10,000), but increases greatly if water percolates through acidic topsoil. Limestone, being rich in calcium carbonate, is slowly dissolved by this solution. The chemical equation for this process is:

The Classical Karst of Slovenia. The River Reka flows through a deep doline before entering the Skocjanske Jame cave system. (Alex Henderson)

Crawl passage formed on a bedding plane in Porth yr Ogof. (Brendan Marris)

$$CaCO_3 + CO_2 + H_2O = Ca(HCO_3)_2 \text{ or}$$

calcium carbonate + carbon dioxide +water = calcium bicarbonate.

An important factor of this equation is that it is reversible, and according to the prevailing conditions will proceed from left to right, or vice versa, until an equilibrium is reached. This results in either the dissolving of limestone or the deposition of speleothems (cave formations), depending upon the concentrations of the component parts.

Stages of Cave Formation

First Stages

Consider a 'new' limestone landscape. It will be, initially, most significantly dissolved and eroded by surface water in the form of conventional rivers and valleys. This is why limestone areas often have a network of dry valleys and gorges created by rivers that now flow underground. When these early rivers were flowing a process began in the flooded system of joints and bedding planes that lay beneath them. Water began to flow along these fissures, gradually dissolving the lime-

Phreatic tube in Dan yr Ogof. (Brendan Marris)

stone, enlarging the channels from narrow cracks into tubular conduits. These first caves are shallow and entirely flooded. There are examples of this type of early cave formation in Britain today. In South Wales, the River Mellte has a flooded system developed below it, which is only explorable by divers, and in the Mendip Hills water leaks away into the bed of the River Mells.

As the cave system developed and enlarged it became gradually able to conduct the entire volume of the surface river. The complete disappearance of the surface river only, at first, occurred during periods of drought but as time progressed the capture of the water became complete. Some British

rivers are still in a transitional state and this can be seen at Easegill in Lancashire, where the surface river only flows during periods of high water.

Later Stages

Over thousands of years cave systems develop, growing longer and deeper until (in a fully mature system) no significant surface water flows on the limestone at any point. Water flowing across other rock strata and meeting limestone will vanish at once by seeping into its bed, tumbling into an open shaft, or flowing directly into a cave entrance. This water will resurge from the

Scallops indicate the speed and direction of water flow.

base of the limestone escarpment at or near the water table.

It is likely that the cave system between sink and resurgence will have undergone several stages of development and that original passages will have been abandoned as the water created new channels at deeper levels. Often the evolution of the cave is affected by events millions of years in the past that have resulted in the deposit of odd beds of shale or sandstone. Such a band of impermeable rock may govern the depth of a system, causing it to extend a great distance horizontally until it encounters a fissure that allows entry to the limestone beds beneath. Water entering Langcliffe Pot in Yorkshire resurges 300m lower in the valley below. Langcliffe Pot extends 3km at a shallow depth suspended on thin beds of sandstone. Somewhere beyond the terminal chokes of the cave, the passage must breach these beds and gain its full depth.

One of the main factors in the development of any cave system is a change in the level of the water table. When water reaches this completely flooded zone it forms 'phreatic' passages. Water tables change, lowering as the landscape is eroded by rivers or glaciers, creating 'fossil' passages.

Phreatic Passages

Phreatic passages form in an entirely flooded zone and have distinct characteristics. Because the water is under pressure, erosion is equal in all directions, frequently resulting in tubular cross-sections. Also, because the water is under an hydraulic head and is not constrained by gravity, flow involves an equal proportion of rising and falling. This enables passages forming in the phreatic (permanently flooded) zone to exploit weaknesses that extend very deep below the local water table. Phreatic caves or passages have the following characteristics:

- Tubular passages
- Complex or maze-like
- Large scallops (indicating slow water flow)
- 'Switch-backing' up and down.

In Wookey Hole, Somerset, water rises from a flooded shaft over 90m deep, while in the Grotte de la Luire, a normally dry system in France that only becomes active in wet weather, the water rises over 500m, completely flooding 18km of passages. These are extreme examples of the 'switch-backing' often observed in phreatic caves where the passages can gently or abruptly slope upwards and downwards.

Regular hollows, known as scallops, are formed by water action. They can indicate both speed and direction of flow. The smaller the scallops the faster the water, and examples a centimetre or two wide are usually associated with rapidly cascading water. Much larger scallops, possibly exceeding 30cm, are often seen high on the walls or roof, and these indicate slow-moving phreatic flow. Scallops are caused by water turbulence within the hollow that causes gradual deepening of the feature. The steepest edge of a scallop is always on the upstream side.

The cave waters are constantly probing the limestone for an easier or more direct route. This is especially true of phreatic passages where every weakness will be exploited by water under pressure. Under these conditions

a single passage is liable to split into a complex network of tubes or enlarged joints. This can lead to the phreatic maze, where, instead of forming distinct, single passages, a huge sponge-like network of interconnecting channels develops. There are examples in Britain, such as Knock Fell Caverns in Cumbria, but this is minor compared to the 183km Optimisticheskaja system in the Ukraine, or the 160km Jewel Cave in South Dakota, USA.

Vadose Passages

When water runs freely along the floor of an air-filled passage, erosion is directed mainly downwards and deep trenches or canyons are cut. This type of passage, formed by water flowing under normal atmospheric pressure, is known as vadose. Often tubular passages of phreatic origin are modified by vadose streams to create a keyhole cross-section. This is a very common type of passage, visible in many caves. Vadose caves or passages have the following characteristics:

- Canyon passages
- Small scallops (indicating fast water flow)
- No 'switch-backing'.

As vadose passages become larger and conduct increasing volumes of water, another process can begin. Grit and pebbles carried by the water mechanically wear and erode the limestone, enlarging the cave by the same process that erodes non-solublerock in surface rivers. This mechanism greatly accelerates the process of cave formation.

Fossil Passages

Any passage no longer conducting the water that formed it, is described as a fossil passage.

OPPOSITE: Large scallops in Ogof Daren Cilau indicate ancient phreatic flow. (Rob Eavis)

This may be a short length or a loop, abandoned because the stream has found an alternative route. This type of feature, known as an oxbow, is quite common in both phreatic and vadose caves. A lowering of the water table will leave phreatic passages either abandoned or subject to vadose flow. Often the system is invaded by opportunist streams that do not relate to the original flow, and so it is possible to have a fossil cave with small active streamways utilizing some sections.

The 'Dip' Factor

Water relies entirely on the crack systems within the limestone to determine its course. One very significant factor is the angle to which the bedding planes have been tilted: the angle of dip. This will virtually always dictate the general direction and gradient of flow. For example, dye placed in water sinking at Tor Hole in the Mendips did not reappear in the spring a few hundred metres down the valley, as expected, but at Cheddar, more than 12km away! Water always follows the geologically easiest route, which can be very long and sometimes extremely devious.

Passages that form at right angles to the dip of the limestone are aligned along the 'strike' of the rock. It is very common for water to run down the dip and gain initial depth before turning and following a bedding plane on the strike. Many cave surveys reveal the main axis of a cave system to be an active stream passage aligned upon the strike with all tributaries entering from the up dip side. Ogof Ffynnon Ddu and Easegill caverns are clear examples, the latter showing one obvious anomaly (the Bull Pot of the Witches' inlet system) that is fault controlled.

Faults

Faults are fractures caused by one rock mass moving in relation to another. They can be

A vadose canyon forming an entrance in the Petar State Park in Brazil. (Faye Litherland)

very significant in the formation of caves because they are major lines of weakness easily exploited by water. Faults can extend for hundreds of metres and run in convenient straight lines that can provide a route for water to run against the prevailing angle of dip. Very large chambers and shafts often form along faults and many systems have examples of this. 'Fault Chamber' is a common name on cave surveys. It is often possible to identify a fault underground by observing fault breccia, a mineralized band formed by the grinding action of two fault blocks as they slide past each other.

Resurgences

The downstream end of the cave system, where the water returns to daylight, is known as a spring, rising or resurgence. In some systems where the geological conditions are favourable the cave may maintain a vadose flow for its entire underground journey, but most systems tend to become phreatic as they near the resurgence. Resurgence caves often drain extensive areas and consequently have large underground rivers and passages. Virtually all the commercialized caves in Britain – familiar names like Wookey Hole, Dan yr Ogof, Peak Cavern and White Scar Cave – are resurgence caves.

These resurgences are ancient and mature with the water rising or emerging from the base of the limestone escarpment, often with older fossil passages exiting above. Other resurgences are less mature, being 'captured' where down-cutting rivers or glaciers have breached passages, allowing leakage or flow to the surface. Cheddar Risings and the Llangatock resurgence in the bed of the Clydach Gorge have almost certainly formed

A keyhole passage in Ogof Draenen.
(Brendan Marris)

this way, and, in each case, a mature phreatic system has been intercepted.

Development Since the Last Ice Age

Until 15,000 years ago Britain was affected by the last Ice Age. Ice covered most of the country and glaciers filled the valleys. The active caves became blocked by glacial debris or frozen mud, and any surface water the conditions allowed was unable to penetrate underground. As the glaciers carved the valleys, deeper existing caves were blocked or truncated. Then, after 50,000 years of Ice Age, the climate changed and a great thaw began. Torrents of water, unable to be conducted by the existing caves, carved deep valleys and gorges. The most spectacular of these in Britain is Cheddar Gorge, which shows the classic features of an immature river valley.

As the melt water cut down, more caves were bisected. Phreatic systems drained as the water table lowered and cave development began again at lower levels. Some existing phreatic systems were so close to the new valley floors that they began to leak upwards to new resurgences in the river bed to create new 'captured' resurgences. Cheddar Risings and the Llangatock resurgence in the Clydach Gorge probably became active during this period.

Few existing cave systems were unaffected by the changes in the landscape. Most British caves now consist of not fully mature active systems interconnecting with ancient preglacial fossil passages. This is well illustrated in the Lancaster Easegill System, where the river vanishes into a series of small sinks but in wet conditions flows in its bed across the entire limestone exposure. Underground, the active vadose streamways intercept old

OPPOSITE: A fine group of stalagmites in Ogof Ffynonn Ddu. (Brendan Marris)

sections of sometimes huge and extensive old phreatic passage. There are, in effect, two cave systems superimposed upon each other.

Speleothems

As previously described, rain absorbs carbon dioxide from the air and then filters slowly through any topsoil, until it penetrates the limestone through the numerous cracks. In the topsoil, it often absorbs significantly more carbon dioxide and becomes much more acidic. As this water filters slowly down through cracks in the limestone the rock is dissolved, and held in solution in the form of calcium bicarbonate.

When this solution emerges into a cave passage it enters an atmosphere with a lower concentration of carbon dioxide. The original chemical equation reverses as the water regains equilibrium by releasing or 'gassing off' carbon dioxide. An effect of this reaction is that calcium carbonate crystallizes out of solution, usually forming the minerals calcite or aragonite, and leaving a deposit on the roof, wall or floor. Other minerals dissolved by the water can give the formations vivid colour: pink or deep red from iron, grey and black from manganese or lead, and a variety of other shades from rarer, diverse minerals.

Cave formations are collectively known as speleothems and can take many shapes and forms on vastly differing scales. Flowstone forms where water trickles down a wall or across a floor. It can assume bizarre shapes similar to melted wax and often forms massive deposits that can completely block a passage. Gour, or rimstone, pools form where water builds its own dam, running in a film over a lip of flowstone. The height of the gour depends on the steepness of the passage and can reach several metres. One of the best examples in Britain is in Saint Cuthbert's Swallet in Somerset where the

The Courtesan, an exquisite formation hidden deep in the great Welsh system of Agen Allwedd. (Brendan Marris)

Great Gour is more than 5m wide and high.

Water running slowly along a sloping roof or overhanging wall often leaves an even, linear deposit. This develops into a band of calcite, often equal in width to one drop of water. As the formation grows it sometimes develops curves that become more and more pronounced as the formation enlarges. The final result is a curtain that can be beautifully curved and folded, often banded by slight changes in the mineral content of the water. The sloping bedding planes of Mendip create ideal conditions for curtains and the best examples in Britain are to be found there.

Flat level ceilings encourage the growth of straws, which are thin tubular stalactites. They are equal in diameter to a single drop of water and illustrate that crystallization is most likely where the water is in contact with air, in this case around the circumference of the drip. Water feeds down the centre of the straw, which can grow to about 3m until finally breaking under its own weight.

Helictites defy the laws of gravity by growing outwards in any direction, often branching and subdividing as they do so. They are generally small (a few centimetres), but the Antlers in Ogof Daren Cilau, South Wales, are nearer to half a metre. There are many theories to explain their origin, ranging from draughts and capillary action to electrical charges. They probably result from a complex combination of factors.

Cave pearls develop when water drips into a pocket containing grains of sand or grit. The grains are agitated while being evenly

Curtain in Saint Cuthbert's Swallet. (Alex Henderson)

Gour pools in Bull Pot of the Witches. (Tom Philips)

coated with a deposit of calcite. They gradually enlarge, often reaching a diameter of over 2cm, until they become too large to move and blend into the calcite deposits around them. False cave pearls are pebbles that have been rounded by constant dripping.

Another type of formation is called moonmilk. This is usually pure white (occasionally light brown or orange) and has a fluffy appearance. It is a soft, chalky deposit most commonly seen coating cave walls and roofs close to the surface. It is a very attractive deposit but is easily damaged by any physical contact, especially by casual visitors leaving graffiti with their fingertips.

Mud formations can assume some very odd and interesting shapes. The most curious ones are formed where water has drained from a pool, leaving complex and almost inexplicable forms behind. Any deposit of mud on the floor of a virgin cave has a pleasantly smooth untrodden look and this should be preserved whenever possible, which means walking in single file and following existing footprints.

There is no fixed rate at which formations grow, despite the figures confidently quoted by showcave guides. Really massive formations may have been growing for tens of thousands of years, but, if the conditions are precisely right, formations can grow surpris-

'Urchin' helictites in Ogof Daren Cilau. (Rob Eavis)

ingly quickly. This is illustrated in some Mendip caves where 2cm thick curtains are growing on guide tapes after only twenty years.

Examples of Speleothems

To see fine formations is one of the great motivations for the caver and there are many systems in Britain renowned for the quality and diversity of their calcite deposits. The great flat roofs of Yorkshire and South Wales have produced some tremendous displays of long straws, while the sloping bedding of the Mendip Caves has created superb flowstones and curtains with warm yellow and orange hues. But pride of place, by popular opinion, goes to Otter Hole in the Forest of Dean. This system begins with long crawls in liquid mud before emerging into a series of fabulously decorated chambers.

The caves on the Continent often have formations on an entirely different scale, well illustrated in the Gouffre Berger's much-photographed Hall of the Thirteen, where a cluster of 10m stalagmites rise from great terraces of huge gour pools. The most beautiful cave in the world must be the incredible Lechuguilla in New Mexico, USA. The strange chemistry of this 138km system (formed by sulphuric acid in phreatic conditions) has led to incredible crystal growths. There are tremendous conventional formations of every type, but almost every rock surface is covered in white prickly clusters of aragonite. Some chambers have huge branching crystal growths that can extend for metres and hang like bizarre chandeliers.

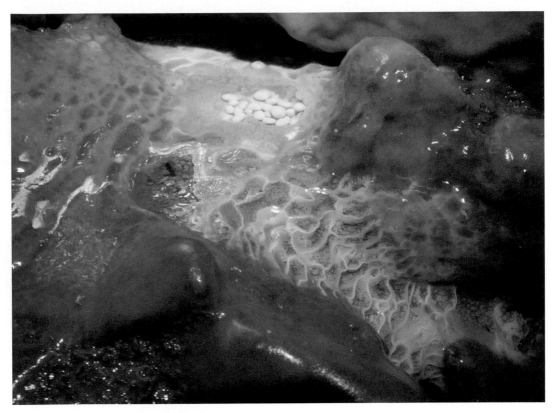

A nest of cave pearls. (Tom Philips)

Pseudo-karst

The name karst is applied to any area showing characteristics of limestone landscape and related cave formation, while the term pseudo-karst is used to describe caves that form in other geological conditions.

The most common non-limestone caves are lava tubes, which are found in volcanic regions such as the Canary Islands and Hawaii. These caves are formed when a lava flow develops a continuous and hard crust, which thickens and forms a roof above the flowing lava stream. Lava tubes can extend, usually as large single passages, for more than 10km.

Kazumura Cave in Hawaii is the longest lava tube in the world at more than 60km long. It is also the deepest cave in the USA and the eighth longest: a trip from the Progress of Man entrance near Volcano National Park to the lowest entrance, in Paradise Park near Hilo, takes two full days and includes seventeen rope drops and 50km of traverse.

Another example of pseudo-karst is the gull cave, which is a fissure, or series of connecting fissures, formed mechanically by slippage of a rock mass. British examples are: Dove's Nest Caves in Cumbria, a multi-entrance complex of underground chimneys; the Windy Pits of Ryedale, North Yorkshire; or the much more horizontally extensive Sally's Rift in Somerset.

Around the world, caves have been recorded in sandstone, conglomerate, salt, granite

ABOVE: Traces of iron give these formations a rusty red colouring.
(Brendan Marris)

OPPOSITE: Pure white flowstone in Peak Cavern. (Brendan Marris)

Part of Kipuka Kanohina lava tube system in Hawaii. (Faye Litherland)

and gypsum. Some of the processes are obscure, but in most cases the rock has a high content of calcium, and the basic chemistry is similar to that which forms limestone caves.

There is a world of fascination in caves for those who want to piece together the visible fragments, to unlock the mysteries of the time capsule. The more we understand the environment into which we venture the greater its capacity to fascinate us. And to the caver who would discover and explore new systems or extensions through digging: the better you understand the mechanisms of cave formation, the greater your chances of success.

CHAPTER 2

Life in Caves

The tenacity of life and its ability to colonize apparently inhospitable environments is well illustrated underground, and often the apparently barren and sterile cave conceals its own tiny ecosystem. The flora and fauna of caves can be divided into three categories. Trogloxenes are occasional visitors to caves, while troglophiles are cave dwellers by choice without special adaptations to the environment. Troglobites live and reproduce underground and often have characteristic adaptations when compared with closely related surface dwellers. Most commonly these are a lack of wings, degenerate or absent eyes, longer appendages, larger or smaller bodies, and partial or total lack of pigmentation.

British Cave Life

The great majority of cave life in Britain falls into the troglophile group, but some have developed subtle differences and might arguably be classed as true, or perhaps evolving, troglobites. This category includes various beetles, millipedes, flies, springtails, flatworms and isopods. One apparently genuine troglobite is a small money spider (*Porrhomma rosenhaueri*), which can be found in just two known caves in Wales. All these tiny creatures are vulnerable to disturbance or trampling by unobservant cavers, and at risk from toxic pollution, which can

originate from the surface, or the contents of a carbide lamp.

The common cave spider (*Meta menardi*) can often be seen in the twilight zone of caves (look up in the roof next time you enter a dry cave) but does not inhabit the dark region. It is quite large, with a brown shiny body and is often seen close to its large white spherical egg case. One Mendip cave was dug into after a caver noticed an egg case beneath a small overhang on the surface. Two species of moth, the Herald and Tissue, spend long inert periods underground and can often be seen, usually quite close to an entrance.

Fish are sometimes seen in British caves, especially trout, which are common in Ingleborough Cave in Yorkshire and Little Neath River cave in Wales. Although these appear white and unpigmented, they are quite ordinary fish that have been carried underground and their normal colouring is quickly restored when exposed to daylight. It has been suggested that the trout in Ingleborough Cave have survived a 100m fall down Gaping Ghyll.

Fungi thrive in the damp dark conditions and quickly exploit any organic material. The spores are either carried in on air currents or attach themselves to cavers' clothing. Any discarded food, or sometimes clothing, develops a white fluffy growth, but the most dramatic examples develop on wooden props in disused mines. Remember that the

A Cuban fruit bat on the wing. (Gina Moseley)

fungi are consuming the wood and it may be dangerously unsafe, so admire the display from a safe distance.

The most obvious and significant trogloxenes are bats. Eleven out of fifteen British species are known to use caves and some, for example the rare Greater Horseshoe Bat, depend entirely on their underground roosts. The Lesser and Greater Horseshoe bats are often seen in their inverted roosting position, suspended from the ceiling by their feet, but other species crawl into crevices and are less conspicuous. Some British bats are on the verge of extinction and others have drastically declined in numbers (the Greater Horseshoe by a factor of 98 per cent). The main cause is probably the change in farming techniques, which has reduced the insect population from which bats feed exclusively. Disturbance is another factor and cavers should ensure that this is minimized. Certain critical sites for endangered species have necessarily been closed during the winter months when the bats are hibernating and most susceptible to disturbance. However, it is notable that bats continue to roost in some very popular begin-

ners' caves, suggesting that some species can happily coexist with cavers, and that overzealous control of bat roosting caves is not always essential.

Bats are protected species under the Wildlife and Countryside Act 1981 and it is an offence to disturb, kill, injure, handle or photograph a bat without a licence. It is also an offence to damage, destroy or obstruct access to any place that a bat uses for shelter or roosting. In order to protect the bat population and to obey the law, the following code should be observed:

- Never handle a bat, and be especially careful not to dislodge them from their roosting position when moving through low passages.
- Do not photograph roosting bats, as flashguns can be very disturbing.
- Bats hibernate from November to April. Be especially careful not to disturb them during this period.
- Do not linger, shine lights on to them, smoke or make excessive noise.
- Do not take large parties near bat roosts in winter and avoid disturbing activities

This isopod, Titanethes albus, in a Slovenian cave, is depigmented and eyeless – a true troglobite. (Alex Henderson)

such as rescue practices or digging.
- Do not use carbide lamps in bat roosts as they cause both heat and fumes.
- Do not use explosives in known bat caves during the winter, or at any time when bats are in the vicinity.
- Leave an access provision for bats when gating a cave, even if no bats have been seen underground.

Overseas Cave Life

Cave life as an international phenomenon is a vast and complex subject that can only be

The eyeless Proteus is found only in a few caves in Slovenia. (Andy Sparrow)

examined superficially here. As more caves are explored around the world and more scientific studies are made, an increasing number of species and self-contained ecosystems are being revealed. While the greatest diversity of cave life is in tropical regions, there are some famous examples in Europe.

The best known of these is the cave salamander *Proteus anguinus*, which inhabits some caves of the Classical Karst in Slovenia. This bizarre creature grows from 20 to 30cm, has no eyes and is flesh coloured. It is able to live for years without eating, an ability that has probably evolved to ensure its survival when marooned in small pools by receding flood water. I have also found an extremely large white spider, complete with hairy legs, crawling over me deep in a Slovenian cave.

Cave biologists have been very excited by a discovery at Movile cave in Romania, which was revealed by accident during construction work. The previously sealed environment, despite having an atmosphere rich in carbon dioxide and sulphur dioxide, was found to support abundant life. Of more than forty species recorded at the site, thirty are new to science. The ecosystem is apparently based on bacteria that form in geothermal waters and are then subject to fungal growth, supporting a food chain that includes snails, water scorpions, leeches and spiders.

Caves in tropical regions frequently teem with life. The basis of the food chain is bats, and sometimes swiftlets, which roost underground in huge numbers. Some caves in Indonesia are estimated to have bat populations of more than 500,000 individuals, which collectively deposit literally tons of guano every day. This nutritious deposit heaves with maggots, millipedes, cockroaches and a host of other creatures, which are themselves preyed upon by larger animals including snakes and porcupines.

CHAPTER 3

Getting Started and Trip Preparation

Learning to Cave

Safe caving must be built on the two foundations of training and experience. If the novice caver is able to learn, underground, in the company of skilled and experienced cavers, these two essentials may be combined. However, some cavers are less skilled and experienced than they like to admit, which means that misinformation and poor techniques can also be disseminated. The novice is wise to gain a second opinion from a modern manual, video or training course.

Introductory caving sessions and training courses are available throughout the various caving regions of Britain. There are various companies, centres and freelance instructors providing either scheduled or bespoke training days or longer courses as required. The appropriate qualification for those instructing technical skills is the Cave Instructors Certificate (CIC) issued by the British Caving Association (BCA).

Formal instructed sessions are very constructive, but they are only a starting point. It is important to take an active interest in techniques, to learn from the experiences of other cavers, to read manufacturers' instructions carefully before using new equipment, and to practise vertical skills on the surface before applying them underground. It must be remembered that many accidents result from the misuse of modern equipment. Most new gear is labelled with the warning 'Training is essential before use', and this does not mean a casual demonstration during a caving trip.

The caver needs experience in a variety of caves – wet and dry, horizontal and vertical – to be fully competent. In this way the skills of basic progression and route finding can be perfected and an awareness of hazards developed. There is another factor that should become apparent: successfully achieving an objective depends on the overall stamina, skill and motivation of all concerned. Human interaction is another vital element underground and dealing with problems like low morale or lack of confidence within the group is an invaluable skill. It is through the caving club that this experience is usually and most effectively gained.

Clubs vary greatly in their aims and attitudes. Some are small and informal while others, particularly those equipped with a cottage or hostel, may have a membership of several hundred. While some clubs organize training courses or encourage participation in training events, others are much less enthusiastic about formal training, which they see as the intrusion of commercialism into their informal and amateur domain. Being a long-established club, though, does not guarantee safe practice. Clubs are made up of individuals, not all of whom may be ideal mentors for the novice, but a few tactful enquiries by a new member should help to identify the safest and most responsible cavers.

The best advice to the novice caver is to find a friendly, welcoming club with a positive attitude towards safety and training. A good club will publish a meets list including the contact telephone number of each trip leader, which enables the novice to check in advance the difficulty, duration and suitability of the planned session. You can find links to caving clubs at various websites, such as www.trycaving.co.uk or www.caving.uk.com. The most popular cavers' forum can be found at http://ukcaving.com and this can be an effective way to contact clubs and cavers or to seek advice and information. It also provides an interesting insight into the caving psyche!

Cavers cannot develop genuine expertise by simply following others on long or difficult trips. Self-sufficiency, judgement and route-finding skills are only learnt by taking an active role in planning and leadership, which is why less experienced cavers benefit enormously by organizing their own explorations of easier caves. The use of surveys and guidebook descriptions, coupled with advice from experienced club members, should enable a group to achieve their own modest but very satisfying explorations. It is this type of activity that complements training in developing the skills, experience and self-reliance of the caver.

Planning a Trip

Information

Any group of cavers planning to visit a particular cave for the first time needs sources of information. The local guidebook is the most convenient initial reference work and includes details of cave locations, access, general layout, route-finding, particular hazards and tackle requirements. Guidebook descriptions are not always easily interpreted, or entirely accurate. They represent one caver's impression of the complexity and hardships of the route and are consequently very subjective. Some area guidebooks are out of print but it is often possible to locate a used copy online (for a comprehensive list of UK guide books see Further Reading below).

There are many other online resources that are useful. The forum at http://ukcaving.com harnesses the collective experience (and opinions) of some very experienced people who are usually happy to advise. 'Googling' a particular cave will usually turn up something useful, such as a logbook trip report, some photographs, or even a detailed and comprehensive description.

Guidebooks generally use one of two grading systems to give some impression of the difficulties that a cave presents. The older system grades begin with Easy, and then progress through Moderate, Difficult, Very Difficult, Severe and Super Severe. These grades are useful guides to inexperienced cavers, but they can be misleading. A short cave might be dangerously unstable but still graded Easy, while another cave with a Severe grade may only have real difficulties in its furthest extremities.

The other grading system uses numerical grades to indicate the following:

Grade I   Easy cave, no pitches or difficulties
Grade II   Moderate cave
Grade III   Cave without particularly difficult, hazardous or strenuous sections
Grade IV   Cave that presents some difficulty, hazard, or large underground pitch
Grade V   Cave that includes very strenuous sections or large underground pitches

Caving magazines and journals can contain very detailed descriptions, photographs and surveys. Some guidebooks include references and many clubs have comprehensive libraries including books, surveys and journals from around the world. The British Cave Research Association administers *Current Titles in Speleology* (CTS), which indexes caving journals. *Descent*, the principal British caving magazine, is also now indexed and often contains useful information (for more information on this magazine visit www.caving.uk.com).

The most useful source of information is often another caver and it has always been an essential function of the caving club to be an exchange of knowledge. Such information is more likely to be exchanged today by email than over the traditional pint. That is

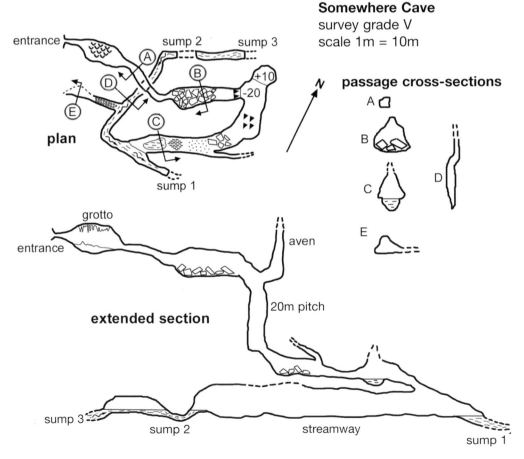

Example of a simple cave survey. Careful comparison of the plan and section creates a three-dimensional impression of the system.

Somewhere Cave

Grade 3
NGR 22446688

Altitude 257m Length 560m Depth 58m

Access controlled by landowner at Bridge Farm (23146785). Goodwill fee payable. Follow the footpath from the farm up the escarpment to stream sink. Entrance is at the base of small outcrop 200m north west from the sink.

The low entrance leads directly to grotto with fine formations. Low passage continues to boulder chamber and head of 20m pitch. Short traverse at head of steep ramp to deviation giving free-hang for lower half pitch. Traverse to left from ramp leads to base of aven, which closes down after 10m. Steep climb at base of pitch leads past short inlet passage on left to sand floored passage and pool. Beyond is climb down to streamway. Danger: streamway liable to severe flooding. Downstream leads to Sump 1, which is choked after 10m. Upstream a fine canyon passage passes low inlet on left and reaches Sump 2. Sump 2 is 8m long (not freedivable). Beyond is short section of low streamway leading to Sump 3, which becomes too tight after 5m.

Example of a typical guide book description giving grade, location and description of the cave.

a shame in some ways since many cavers like nothing better than to swap anecdotes in the pub. Such stories, of course, are liable to dramatic embellishment, which creates another problem for the novice – sorting out what's really true! Good advice can usually be obtained from caving shops, and most regions have at least one. Professional instructors are also usually happy to give advice when asked.

Guidebooks include simple plans or sections of most major caves. While these are useful, however, they are no substitute for a detailed high grade survey. It is the cave survey that is the caver's most essential resource rather than guidebook text descriptions. For long or complex systems, such as Ogof Ffynnon Ddu or Easegill Caverns, the cave survey is an absolute essential to ensure that objectives are reached without becoming hopelessly confused. Unfortunately getting hold of a survey is not always easy as many are out of print and are under copyright. Trying to track down and obtain a copy of a particular cave survey can be both time-consuming and frustrating even with the benefit of the internet. This is one area where the caving club with a comprehensive library has a clear advantage.

Choosing a Group

Consider the most appropriate group size for the type of trip you are planning. Four is generally reckoned to be an ideal number, since it's a large enough group to deal with emergencies but small enough to make reasonable progress. At one time four was stipulated as the safe minimum but experience has shown that imposing a mandatory group size can compromise safety in other ways. If the trip is particularly demanding, or if it requires a very high standard of SRT (single rope techniques) proficiency, it may be difficult to put together a suitably experi-

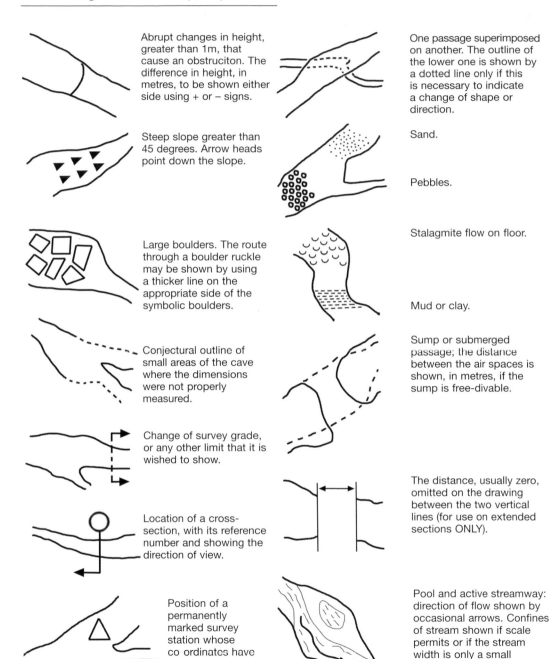

Abrupt changes in height, greater than 1m, that cause an obstruciton. The difference in height, in metres, to be shown either side using + or − signs.

One passage superimposed on another. The outline of the lower one is shown by a dotted line only if this is necessary to indicate a change of shape or direction.

Steep slope greater than 45 degrees. Arrow heads point down the slope.

Sand.

Pebbles.

Large boulders. The route through a boulder ruckle may be shown by using a thicker line on the appropriate side of the symbolic boulders.

Stalagmite flow on floor.

Mud or clay.

Conjectural outline of small areas of the cave where the dimensions were not properly measured.

Sump or submerged passage; the distance between the air spaces is shown, in metres, if the sump is free-divable.

Change of survey grade, or any other limit that it is wished to show.

The distance, usually zero, omitted on the drawing between the two vertical lines (for use on extended sections ONLY).

Location of a cross-section, with its reference number and showing the direction of view.

Position of a permanently marked survey station whose co ordinatcs havc been published; not shown if it would obscure other detail.

Pool and active streamway: direction of flow shown by occasional arrows. Confines of stream shown if scale permits or if the stream width is only a small proportion of the total passage width.

Survey symbols.

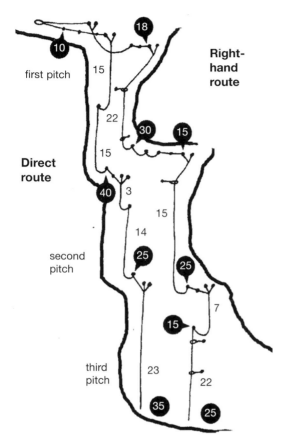

Right-hand route

Direct route

first pitch

10

15

18

22

30 15

15

40 3

15

14

second pitch

25

25

15

7

third pitch

23

22

35 25

A rigging topo gives precise information on rope requirements and anchor points.

more relaxed club caving world of today there is often not a designated group leader and things are quite informal. Collective leadership does not always work to everyone's benefit. For example, it may be apparent that the novice sandwiched amongst experienced cavers is struggling, but there is sometimes a reluctance to take responsibility and to abort or modify the objectives of the trip. This is a scenario that sometimes leads to cave rescue, but more often to the novice having a miserable time.

Equipment

Tackle requirements often vary from the guidebook listing and it is usually advisable to carry a short additional rope in case members of the party do not agree with the guidebook writer's definition of 'easily free-climbable'. Pitch tackle requirements generally list the required length of ladder and lifeline, but do not allow accurately for rigging ropes or traverse lines. If the cave is being rigged for SRT, additional information such as the length of traverse lines and the position of rebelays and deviations is very useful. SRT guidebooks are available in most regions and contain 'topo' surveys of selected caves with all relevant rope requirements and distances listed.

The group should be equipped for problems and emergencies. It is the responsibility of each caver to ensure that their personal equipment is well maintained and that they have adequate back-up lighting. The group, or leader, may decide that one spare light per person is unnecessary, but remember that if all the lamps have been charged together on one unit they can all fail together, too! Personal or group first aid and emergency kits are essential and easily assembled (see Chapter 9). It's very wise to get into the habit of carrying a fleece balaclava and a lightweight survival bag. Some caves are a long

enced group. The team can only be as strong, as fast or as proficient as its least experienced, confident or physically able member. In a system with deep pitches a small group is always preferable to reduce delays in a cold and hostile environment. Diccan Pot, for example, with its powerful waterfalls, chilling draughts and technical rigging, is far better suited to an efficient party of two than a mixed group of four.

Choosing a group for a particular cave is an important responsibility. In the old, post-war days the party had a leader who was empowered to make all essential decisions, including the selection of the team. In the way from the nearest road (often more than 3km in Britain). The walk is often across attractive open country and should be enjoyed rather than endured. It helps to carry kit in a comfortable rucksack and you will find a lightweight waterproof jacket more comfortable to walk in than an over-suit. In the depths of winter a hat and warm gloves are advisable, if not for the walk up, then at least in readiness for a cold trudge back in adverse weather.

Food, Energy, Alcohol and Water

The human body has much in common with an internal combustion engine. It must be fuelled before use and then topped up again as required. It also functions most effi-ciently when run economically: in other words, not in bursts of speed but at a steady, sustainable pace. The body must also fuel its own heating system, and the more energy that needs to be diverted to this function, the less that remains to move through the cave – a correctly dressed caver is likely to have more endurance than a poorly dressed caver.

A high energy meal such as the traditional English breakfast or a pasta dish is impor-tant before going underground and is best consumed one to two hours before entering the cave. Underground, energy levels can be maintained by chocolate bars or by glucose drinks, which have the benefit of hydration. It can feel slightly ridiculous carrying water to a cave that already has water in abun-dance, but many cave streams are badly pol-luted so you need to think twice before drinking from them. On a long arduous dry trip you are likely to become very dehydrat-ed if you don't pack that water bottle.

A word about alcohol: this has several negative effects on the metabolism that last for up to twelve hours after consumption. It causes dilation of the surface blood vessels, which increases the danger and effects of

hypothermia. Watch out for that hung-over caver who can't face breakfast but is still determined to come on the long trip.

Weather Forecast

In view of the fact that flooding is one of the single greatest causes of rescue and that most caves are affected to some extent by high water, it stands to reason that we need to assess the risk before venturing under-ground. Many guidebook descriptions include warnings of caves, or sections of caves, known to be floodable. The omission of a warning should never be interpreted as a guarantee of safety and this is another area where advice from local cavers is invaluable. If there is any risk of water levels affecting the cave it is essential to get an accurate and up-to-date weather forecast from a reliable source, and to assess the current conditions (for more on hazards see Chapter 8).

Emergency Call-out

Before setting off, an emergency call-out must be arranged and should be left in writing with a responsible person, preferably a caver who understands cave rescue call-out procedures. The more information they have, the easier it is for cave rescue to assist in an emergency.

The minimum information required is:

- Date of visit
- Leader's name, address and telephone number
- Number in party
- Car registration numbers
- Cave to be visited
- Estimated time in
- Estimated time out
- Rescue call-out time. If the cave can be approached from more than one direc-tion, the location of car parking and the

proposed surface route is required. If the cave system is extensive, the proposed route underground should be recorded in detail.

Access

It is important to ensure that any access agreements and procedures are adhered to. Guidebooks contain this information, but it is liable to change at short notice. The BCA website (www.british-caving.org.uk) contains links to the various regional caving councils of Britain. It is a good idea to refer to these sites to double-check that access arrangements have not changed. You will find other useful advice and information pertaining to the regional caving areas too. If the landowner does not give permission to cross the land or descend the cave, the decision must be accepted quietly and politely.

One of the greatest threats to access is bad behaviour on the surface and it is particularly important that groups visiting caving areas conduct themselves responsibly. The following code should always be observed:

- Park sensibly (not obstructing gateways or any right of way)
- Leave no litter
- Change discreetly, without causing offence
- Keep to footpaths
- Minimize rowdy behaviour (especially late at night)
- Observe the country code.

The Right to Roam

The Countryside and Rights of Way Act 2000 (CRoW) created the right to roam on designated areas of upland Britain. You will be delighted to know that you have the right to walk to many previously restricted caves and 'to enter and remain on any access land for the purposes of open-air recreation'. However, you may not bathe 'in any non-tidal water', or

obstruct 'the flow of any drain or watercourse'. Until such time as the law is tested it remains unclear how the Act affects our access to caves. There are already reports of cave entrances being blocked by landowners in access areas and it may be that the CroW Act ultimately does us no favours whatsoever.

Vehicle Security

It is a depressing fact that every year literally hundreds of cavers are affected by thefts from cars. Cavers make lucrative and easy targets because they are often obliged to leave their clothing and valuables in vehicles for long periods in remote spots. There are ways to reduce the hazard, such as parking in a more secure area, even if this involves a longer walk. Whenever possible, leave nothing visible in the car to attract the thieves and be on the look out for suspicious behaviour. This problem is not restricted to Britain and plagues cavers in many other countries.

Surface Navigation

Having selected your cave, chosen the team, planned the route and arranged a call-out, it now remains to find the cave entrance. If this is a gaping hole next to the road there is no problem, but a small entrance on a misty moor is problematic. A map and compass are essential and their usefulness is greatly enhanced by some basic navigation skills. It can be easy to find the way to an entrance in daylight, but when the party emerges, hours later in the dark, it may be a different story on the way back. Often when the weather deteriorates the moor or mountainside can be a more hostile environment than the cave.

A 1:25,000 scale map is most useful along with a Silva-type compass that can be used to take bearings. Use the map to prevent getting lost in the first place, which means knowing exactly where you are starting from, and what the route will be. The magnetic variation must be taken into account, the degree of which should be indicated on the map legend. Remember: 'grid to mag add' – when taking bearings from the map add on the variation, and 'mag to grid get rid' – deduct the variation if transferring the other way. You can only use the compass effectively by taking bearings on distant and continuously visible objects, not simply by walking in the general direction indicated.

Most walkers average 5km/h or 1km every twelve minutes on level ground, but one minute per every 10m height gain, or one contour line, should be added. By timing the walk it should be apparent when you are in the right area. If aiming for a specific point on a linear feature, such as a junction of walls, it is best to 'aim off' and hit the feature to the right or left. It should then be a certainty which direction the feature should be followed to reach the correct point.

It is useful to know how many of your paces equal 100m, especially when looking for a small entrance in misty conditions. Navigate to an obvious nearby feature and use this as an 'attack point'. Count paces, and follow bearings so that you can always find a way back to the attack point if the first search is unsuccessful.

There is, of course, an alternative to the traditional skills of map reading and navigation, which is to use GPS. The Global Positioning System, which computes a precise position from a satellite network, has become a practical and affordable option for cavers. Using this system you simply input the coordinates or grid reference and the device will direct you to within a few metres of the target. This may be when you realize, on a remote foggy moor, that the grid reference was originally recorded using less precise instrumentation. GPS is especially useful on expeditions to log and locate cave entrances. Quite often, before the days of GPS, in jungle or on *lapiaz*, promising entrances were discovered, noted for exploration, and then never found again!

CHAPTER 4

Personal Equipment

The Cave Environment

Air Temperature

The air temperature underground will reflect the average temperature for the latitude and altitude of the cave. In Britain our average temperature is about 10°C and this is what we find underground. If we go to warmer climates we find warmer caves and in tropical areas we can expect a pleasant 20°C (68°F). The further north we go, or the higher the altitude, the cooler the cave air will be. High altitude caves often remain below 0°C (32°F) and contain spectacular flows of ice, such as the Grotte Casteret in the French Pyrenees and the famous showcaves of Eisreisenwelt and Dachsteineishohle in the Austrian Alps.

Draughts

Caves are well known for maintaining a constant temperature but this does not take into account the significant effects of draughts. Draughts can make caves seriously cold places! There are four main mechanisms that can, often in combination, create draughts underground.

1. Surface Air Pressure

The cave is a vessel containing air, and as surface air pressure changes, rising or falling in response to low or high pressure weather systems, the cave will draught in, or out, as pressure equalizes to match that of the surface. Generally this is a subtle phenomenon but there are exceptions when a particularly massive system has only one entrance. It is not uncommon for a 50km/h wind to blow through the entrance of the huge Lechuguilla system in New Mexico.

2. Temperature

The cave is a reservoir of air at a fixed temperature, which is usually able to circulate through a much greater area than that accessible to cavers by penetrating through boulder chokes and impassably small passages. The mass of air, being relatively warm in the winter and cold in the summer, will sink and rise accordingly, generating significant draughts that seasonally change direction. The greater the volume of air and the higher the differential between underground and surface temperatures the more powerful this type of draught will be. The great cave system of Pierre Saint Martin in the Pyrenees contains a series of massive chambers and during the summer months the lower entrance emits a hurricane of cold air sinking through the cave system. During the winter the warm air rises, sometimes creating telltale blowholes in any snow covering.

3. Blowhole

Caves with multiple entrances that allow a free passage of air often draught vigorously. This is especially true in active systems where falling water generates blowhole draughts by drawing air in from one end of the system and expelling it from the other. In the system of Gaping Ghyll, the complex route through ancient and abandoned tunnels to the Main Chamber and its mighty waterfalls is easily indicated by the gusts of air blowing from that direction. This mechanism is so powerful that it can dominate the other draught mechanisms and remain constant regardless of surface temperature and pressure. It can have a profound effect in the winter, when cold surface air is drawn into the cave.

4. Water Turbulence

This type of draught will be found in the vicinity of any falling water and is caused by displacement and consequent turbulence of the air. It is a very local but extreme effect, which can produce the equivalent of a buffeting squally gale. If the turbulence draught is combined with the blowhole effect in midwinter, when cold surface air is drawn into the cave, the constant $9-11\,^{\circ}$C of the British cave reduces to an effective cooling rate of $-9\,^{\circ}$C ($16\,^{\circ}$F)!

Clothing

The air temperature in caves is highly variable, even in the temperate climate of the British Isles. Regulating body heat is never easy and extremes of hot and cold can be experienced even on short visits to minor cave systems, when strenuous dry sections alternate with deep pools and waterfalls.

Dry Caves

In a genuinely dry British cave, averaging $8-10\,^{\circ}$C ($46-50\,^{\circ}$F), non-specialist clothing is suitable. Several thin layers (T-shirt, sweatshirt and boilersuit) should provide adequate insulation while the wearer is stationary without causing excessive overheating in the more physical sections. Clothing should be loose fitting, especially the trousers, to allow freedom of movement; a boilersuit or overalls helps to keep everything together at the waist. Clothing like this will be quite adequate for the novice initiated in caves like Mendip's Goatchurch or the dusty-dry systems of the Forest of Dean, but not in any situation where they are likely to get wet.

Wet Caves

Beginners, whose initial experience is in a dry cave where they exerted themselves considerably, often form a false impression that the cave environment is quite warm. They are likely to shed much of the warm clothing they were originally urged to wear when they embark upon their next cave and, if this is wet, they may have a cold and miserable time. The benefits of a dedicated clothing system are soon appreciated in an environment that involves saturation in very cold water, often combined with chilling draughts.

General Principles of Clothing

The extent of heat conduction depends upon the medium through which the heat transfers. If the body is wrapped in saturated clothing, much energy is expended in trying to warm a considerable volume of water, which is likely to be cooled, or flushed away and replaced with cold water at every pool or waterfall. Air is much easier to warm than water, and is a much more efficient insulator, which means that materials that drain, and do not remain waterlogged, require less energy to heat. Wool functions in this way to some extent, which is why, in the pioneer days of caving and mountaineering, wool next to the skin was almost obligatory. Today we have 'technical' clothing systems for use in both environments, which are based on the principle of three layers. These are:

Base layer. The innermost layer of clothing, or the layer of clothing worn directly next to the skin, made from a thin 'wicking' material that conveys moisture away from the skin. This may be a one-piece, full body garment or two-piece composite.

Insulating mid-layer. The mid-layer is the crucial insulating layer usually made from

The full wetsuit is only necessary for trips that require prolonged immersion or involve swimming. (Brendan Marris)

synthetic fleece, which drains and dries rapidly after saturation to maintain both warmth and comfort. A one-piece, purpose-made garment, the 'fleecy-suit', is usually used for caving.

Shell layer. Usually a purpose-designed, one-piece 'oversuit' of proofed nylon or PVC. This layer traps air, improving insulation, while protecting from draughts and showering water. It also acts as a robust abrasion-resistant layer to help protect both the under-layers and the caver within. PVC oversuits are ideal for very wet caves but can cause overheating in dry systems. Proofed nylon oversuits are comfortable for most situations that don't involve a waterfall beating down on your head.

Modern materials drain quickly after saturation, providing effective insulation even after a complete immersion. (Brendan Marris)

Wetsuits

Wetsuits were a revolution for cavers when they were introduced back in the 1960s but have largely fallen from favour today. They can still be the best option for trips that require prolonged or repeated periods of immersion or the navigation of deep water. In the 'old' days most cavers used 4–6mm neoprene, usually without any oversuit or outer layer. This may have been adequate, but only barely, and it was always hard to find a comfortable balance – moving too fast in dry passages caused heat-exhaustion, while only a few minutes of inactivity could lead to teeth-chattering cold. Coupled with this was the problem, especially for those choosing 6mm neoprene, that it could be hard work just bending a limb.

Many cavers today prefer a thinner 3mm wetsuit worn under an oversuit. A wetsuit of this type can be bought off the peg in a supermarket and is reasonably effective if it is combined with the oversuit to provide that essential 'shell' layer.

A better option is the 'neo-fleece', which is a 3mm 'shorty' wetsuit with fleece arms and legs. This provides a neoprene layer around the body without the restricted limb movements associated with full wetsuits. Another great benefit of the neo-fleece is the option of wearing a base layer underneath it to provide more comfort and insulation. With an oversuit on top of this combination you have the three-layer system but with the benefit of the neoprene around the body. This is my preferred option for very wet trips.

Accessories

If the feet are likely to stay dry, one or two pairs of thick woollen socks are most comfortable, but if prolonged wetting seems likely, neoprene (wetsuit) socks are advisable. These ensure that the feet stay warm even if they do end up as wrinkly as prunes. Long neoprene socks that reach up to just below the knee are made especially for cavers, and these have the added benefit of protecting the shins.

Gloves help to keep the hands warm and the gauntlet type can prevent water running down your sleeve. They also prevent abrasions to the hands, which is useful when digging and important in caves where a risk of Leptospirosis infection exists. Washing-up gloves do not impair dexterity, but have a short life, sometimes less than one caving trip. Heavier duty, rubberized canvas or cotton gloves are more suitable and longer lasting. They can be modified by punching a hole, which allows them to be clipped to a karabiner while not in use, for instance while rigging.

A fleece balaclava can easily be tucked inside the oversuit and, especially in wet systems, should be a standard piece of kit, carried by every caver: 30 per cent of heat loss is through the head and any measure that reduces this has a real benefit. Remember that the balaclava is not principally to keep your head warm but to prevent the heat loss that is reducing your core temperature and making you shiver. The balaclava should be worn whenever the caver starts to feel cold, and not just saved for extreme situations.

Pads

Protect your knees! They are not designed for crawling and are easily damaged. Two types of knee-pad are available: neoprene and strap-on types used by miners or builders. Neoprene pads provide reasonable protection for a large area, and generally stay

OPPOSITE: Caves in northern latitudes or at high altitude are often permanently frozen. (Peter and Ann Bosted)

Tropical caves are sufficiently warm that specialist clothing is often unnecessary. (Brendan Marris)

in position quite well. Thicker strap-on types provide less protection for the top of the knee and tend to rotate out of position, but they are excellent for long and punishing crawls.

Elbow pads can be used, but it really is advisable to get into the habit of simply not using the elbows at all. Like the knees they are easily damaged, but since it is just as easy to rest weight on the forearms, it is quite possible to avoid using them completely. If you bruise easily and your forearms are black and blue after every trip, try cutting some pads from a closed-cell sleeping mat and hold them in position using a tubular elasticated bandage.

Footwear

Footwear should have a deep tread, support the ankle and be hard wearing. Walking boots with lace hooks may be used for horizontal caving, but never when climbing ladders is required, as the hooks can catch and snag on the wire sides. Caving is very hard on boots and most cavers opt for the simplest and cheapest option – the wellington boot. A suitable pair should be snug fitting and have a good tread pattern. With a little ingenuity, a sealing system can be devised using a strip of plastic cut from an old container and glued or riveted into a ring that fits snugly into the top of a wellie boot. This keeps the boot rigid, and allows the oversuit to be sealed with rubber bands cut from old tyre inner tube. Using this system, it is possible to wade in thigh-deep water and keep (relatively) dry feet.

Helmets

The most obvious, routine function of the helmet is to protect the head against minor scrapes and collisions with the cave roof, but it also guards against other hazards such as falling rocks and provides vital protection in the event of a fall. There is no official specification for caving helmets but we can refer to existing standards applied to similar activities. The caver is subject to exactly the same hazards as the mountaineer and is strongly advised to use a helmet CE marked to standard EN 12492. At the present time no manufacturer is producing a dedicated caving helmet and we have no option other than to consider the available range of climbing and mountaineering products.

Soft items such as survival bags or chocolate bars are not likely to be a hazard if carried inside the helmet, but hard objects certainly will be. Helmets should be replaced whenever they show any signs of damage or excessive wear.

These are some of the models currently on the market:

Petzl Elios. A popular ultra-lightweight climbing helmet. The Petzl Spelios version of this helmet comes with pre-attached Petzl Duo lighting system. Adjustment is by a single rotating wheel and the helmet comes in two sizes, the smaller of which is suitable for young children. These are great products, but

The Petzl Spelios, an ultra-lightweight helmet with integral Duo lighting system.

The Petzl Ecrin Roc, a sturdy, comfortable helmet with easy adjustment.

be aware that the main protective element is the polystyrene inner casing and that the outer shell can split under low impacts, for example by dropping it onto a hard surface.

Petzl Ecrin Roc. Excellent helmets of more conventional construction. The Ecrin Roc has two adjustment wheels allowing a very precise and comfortable fit. This helmet also has very generous ventilation holes, which, in my opinion, are a real benefit for a caving helmet. It can be purchased with a pre-attached Petzl Duo lighting system. This helmet does not really adjust down small enough to be suitable for younger children.

Edelrid Ultra-light. A lightweight, well-ventilated, one-size-fits-all helmet that would be my current recommendation for instructors and outdoor centres.

Lamp Brackets and Helmet Modification

In order to use some caving lights it will be necessary to modify the helmet by drilling some small holes. However, you might be deterred from doing this by dire warnings such as this, taken from one supplier's website: 'If a helmet is modified in any way after manufacture (i.e., by drilling the shell or fitting a lamp bracket), the CE, UIAA, Industrial or other standard is cancelled.' It is difficult to reconcile this statement with the fact that Petzl actually provides a hole-drilling template to allow the fixing of a Duo light onto an Ecrin helmet. This really only becomes an issue in the event of an accident in which the helmet fails at the point of modification. Such failure is unlikely if the holes drilled are as few and small as possible, and if any screws, bolts or rivets are inserted with their flattest side on the inside of the helmet.

Lighting

Caves are entirely and perpetually dark and our lamps are a lifeline, without which we have little chance of returning to the surface. They must be reliable, long-lasting and sufficiently durable to survive immersions, mud and harsh treatment. But our lamps do much more than simply light the way: they reveal the size and extent of caverns; illuminate formations hanging high above us; and show the varied colours of minerals and flowstones. The brighter the light, the more we can appreciate and enjoy the environment that surrounds us.

We are in the midst of a technological revolution as year by year LED lights grow brighter and new battery technology powers them for longer. The choice of products is changing so rapidly that any recommendations made here will be rapidly overtaken by new models and products. The future for cavers is certainly brighter!

Having said that, the old dependable, the miners' lamp, is still used and favoured by some. There are several different makes but the design is identical: a heavy, very robust battery pack that threads on to a belt and supplies a headset on the end of a 1.3m lead. The headset is water resistant and contains two bulbs, a main beam and a reserve, or pilot, light. The unit gives about twelve hours' light on the main bulb and double this on the pilot. A purpose-made (constant voltage) charging unit is required, into which the headset plugs directly.

Miners' lamps are mainly lead-acid batteries and although most are sealed and maintenance free, a few older models contain a liquid electrolyte that must be topped up periodically. Some top-up holes are plugged and sealed while others are simply open vents in the side of the lamp. This latter type causes problems in two respects: cave water gets in and dilutes the electrolyte, and acid escapes to cause skin burns, eat away at clothing or cause dangerous damage to webbing and ropes. To get the maximum life from a lead-acid lamp it should not be allowed to run flat, and never left in a discharged state.

You are now unlikely to encounter the old NiFe-cell with its distinctive metal case, but if you should, beware! This type contains a particularly aggressive alkali, prone to leaking, which causes serious burns. I can testify to this personally, having had both buttocks excruciatingly neutralized with acid after such an episode.

The first purpose-made caving lights, which appeared in the early 1980s, were based on the miners' lamp design but used sealed NiCad (nickel-cadmium) batteries, which are maintenance free and can be left discharged for long periods. The 'F' cell became the basis for several purpose-made caving lamps, most notably the Speleo Technics models, which are still being manufactured today. The two-cell (FX2) battery is relatively light and compact and gives up to ten hours of adequate, if not brilliant, light. The larger three-cell (FX3) version is far better for the serious caver, but for those who

don't mind a really bulky battery pack there is the five-cell FX5, which performs superbly and can even run a 10-watt bulb for four hours (this is the only electric light I have used that doubles as a hand-warmer).

Caving with a heavy battery hanging off a belt and a cable that catches on every protrusion is not ideal and throughout the 1990s many cavers experimented with helmet-mounted batteries. One option was to use a miners' lamp headset powered by a disposable 'flat' battery, a system which some cavers still use. This gives a good light for a few hours but the brightness then diminishes. Second-hand headsets are currently (with the demise of the mining industry) cheap and plentiful, making this initially an inexpensive system, although the cost of the batteries does add up.

A popular helmet-mounted battery light, made by Speleo Technics, is the Headlite. This comprises a compact battery that straps on to the back of the helmet, supplying a headset based on an original miners' design. The battery connects with a simple plug, allowing easy replacement underground. I find this system a bit heavy compared to modern ultra-lightweight lights but its robustness and reliability are well established, making it one of the most popular lighting systems in the UK.

Some cavers like a diffuse spread of light, but most prefer a concentrated, focused beam. The standard miners' headsets usually allow the bulb to be screwed slightly in or out, which adjusts the focus. Some bulbs have no screw thread and require a slightly different reflector. These are pre-focused and give a very concentrated pencil-beam. Halogen bulbs are brighter, but if they do not focus correctly they provide no more light than a cheaper, conventional bulb. The condition of the reflector is very important: if this is dull or corroded the light output will be greatly diminished. The lens, too, needs only a few small scratches to have a similar effect. To get the best from a conventional electric lamp these items need to be replaced regularly.

LED Lights

The majority of cavers now enjoy the reliability, longevity and brightness of the Light Emitting Diode (LED). There is (quite literally) a dazzling array of new products available to us, ranging from 'under a tenner' head-torches, suitable for a dry novice trip, to the ultra high-tech floodlight that is the Scurion. At the time of writing I particularly like the amazingly bright, reliable and lightweight Tikka XP and Myo, which are manufactured by Petzl. Another extremely popular Petzl product is the Duo, so called because it includes both a halogen bulb and an LED array. The Duo is fully waterproof, runs on either AA batteries or a rechargeable NiCad unit, and can either be fixed permanently to a helmet or secured with elasticated straps.

Users of older-style headsets can enjoy the benefits of LED lighting too. Speleo Technics manufactures a combined reflector and LED array for use with its lighting systems and there are other diverse products designed to fit within a miners' headset.

Carbide

An alternative to electric lighting is the carbide lamp. Calcium carbide looks similar to limestone chippings, but reacts with water (with a lot of fizzing and bubbling) to produce acetylene. The carbide lamp is a simple generator consisting of two chambers. Water, controlled by a needle valve, drips from the upper chamber on to the carbide below, which gases and supplies a burning jet, where it is ignited and burns with a brilliant white flame.

There was a time, before the advent of

Speleo Technics Nova, a powerful LED spot, waterproof to –50m.

The superbly engineered headpiece of the Scurion with two ultra-bright LEDs.

Petzl Duo, currently the most popular caving lamp in Britain.

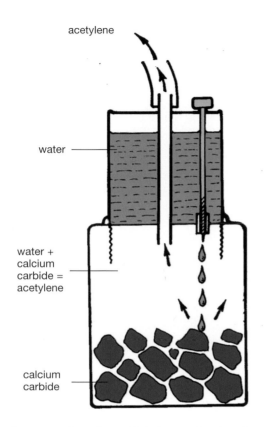

acetylene

water

water +
calcium
carbide =
acetylene

calcium
carbide

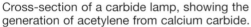

Cross-section of a carbide lamp, showing the generation of acetylene from calcium carbide.

The Petzl Explorer lighting system, which combines carbide and electric lighting.

reliable electric lights, when these acetylene-powered brass lamps were standard kit. The helmet-mounted 'stinky' was ubiquitous, as were the special skills required to keep these notoriously fickle lamps ignited. With every splash of water and bash against the roof extinguishing the flame, and with only a flint-wheel vainly operated by wet hands, relighting could be a frustrating process. I recall being dispatched to search for a group of ten missing cavers and finding the entire party marooned in total darkness. Such events were hardly surprising considering that the lamp would only operate when topped up with water every twenty minutes or so, and that a full charge of carbide seldom lasted more than a couple of hours.

Carbide use in the era of the 'stinky' had a profoundly negative effect on the cave environment, as the short duration of the lamp led to the spent residue, a toxic grey paste, being frequently dumped underground. This is one reason why carbide was banned from use in a number of cave systems.

Inevitably the 'stinky' fell out of use but the carbide lamp went through a considerable revival during the 1980s and 1990s. A new generation of lamps was produced using much bigger, waist-slung generators, supplying a helmet burner with a hose. These lamps incorporated a piezo igniter and an integrated electric back-up, and would last for most trips on just one fill of carbide.

The Petzl Tikka XP makes an excellent back-up light.

I went through a period of using this type of lamp and greatly enjoyed the diffuse pool of warm yellow light it created. There was another reason why these lamps were favoured in what might seem the most inappropriate situations – wet and windy vertical caves. The reaction between the water and carbide is exothermic and produces heat, allowing the generator to be used as a very effective hand-warmer, or even body-warmer when slipped down inside the oversuit.

Carbide is best transported underground in a 'carbide banana' made from a length of old inner tube, which can also be used for taking spent carbide out of the cave. In the pre-LED days, groups visiting remote areas where recharging batteries was not an option usually relied on carbide as the only viable lighting system. The efficiency of modern lighting systems means that a caving expedition today is more likely to use a pack of AA batteries than the traditional drum of carbide. A cautionary tale from back in the days when cavers used carbide and favoured

ammunition boxes to carry their goodies. A few people made the mistake of putting their spent carbide in one of these airtight containers. The carbide, however, was still producing enough gas to fill the box to a pressure of two atmospheres, at which point it spontaneously explodes! The result was several very distorted ammo boxes and a number of traumatized, but otherwise unhurt, cavers.

Spare Lighting

Every self-sufficient caver needs a back-up light for that inevitable occasion when the main light source fails. One option is to fix a small waterproof torch to the side of the helmet, which is easily arranged with an elasticated loop if the helmet has ventilation holes. My preferred back-up light at the moment has to be the Petzl Tikka XP, which can be attached to the helmet or hung around the neck. There is a good argument for not having your spare light attached to your helmet, as a few cavers can testify – that's when you forget to do up your chin-strap and your helmet, light and spare, descend a pitch without you!

Leaders of organized groups such as outdoor centres or university clubs often feel it unnecessary to apply the 'spare light for every caver' rule, but those of us who work in instructed caving inevitably experience the day someone forgets to charge the lamps. Half an hour into the cave, when you find yourself confronted with an array of fading lights, you realize it's going to be a slow journey out on that one dim old headtorch you chucked in the bag…

OPPOSITE: Wet and windy conditions may be a problem for carbide users but it can provide a welcome heat source. (Chris Howes)

CHAPTER 5

Horizontal Techniques

Between the tapes and on the knees is the best way to preserve the straws and mud floors in Ogof Draenen. (Brendan Marris)

It is a rare luxury to be able to walk upright on an even, level floor for any distance underground. The route is more likely to involve a succession of crawls, stoops, wriggles, squeezes, slopes, climbs, chimneys, pools and canals. Each obstacle, every bend and twist of the passage, must be negotiated without expending unnecessary energy. Good cavers are not always super-fit, but they are very energy-efficient. An expert caver eases his or her way through the cave, without the thrashing or struggling that often leaves novice cavers bruised, battered and exhausted. The ability to move efficiently through a cave is not a skill that can be easily learnt or taught, as it depends very much on natural aptitude. Having said that, enthusiasm is the most essential component, and with sufficient tenacity and practice most devotees develop a reasonably efficient technique (eventually!).

Conservation

It is impossible to explore a cave without having some impact upon its environment, but by adopting a responsible attitude and observing a simple code of conduct this can be minimized. The caver should always try to move with care and appreciation; there is no need to walk upon untrodden surfaces when an existing path can be followed, and no need to lay a muddy hand on virgin calcite when by a little extra effort it can be avoided. Much of the damage that occurs underground is caused by casual or inexperienced visitors who have not been educated in the Conservation Code:

- Do not touch formations
- Observe marker tapes (this includes photographers!)
- Leave no litter, spent carbide or human waste
- Do not smoke underground
- Do not mark or write on cave walls
- Avoid damage to original floors by following established paths
- Do not disturb bats or any other cave life
- Move slowly and carefully in decorated areas
- Keep novice parties at a manageable size
- Ensure that novices are educated in this code.

Walking and Stooping

Walking passages present no difficulties and it is all too easy to rush on at speed. In doing this the caver may not notice the fine formations high in the roof, the side passage concealed at floor level or the mud formations they are happily tramping over! Caving is a marathon, not a sprint; movement should be at a steady measured pace so that the surroundings can be appreciated. Many broken formations are at head height; with more care and less speed from passing cavers they might still remain intact.

As the roof lowers, stooping becomes necessary. Long sections of stooping can be very uncomfortable, especially when it is com-

Deliberate vandalism or carelessness – the effect of either is permanent. (Chris Howes)

Flat-out crawling over cobbles in Cwm Dwr. (Brendan Marris)

bined with an awkward sideways walk, such as is needed for Stagger Passage in Langcliffe Pot. A sort of gibbon action can be perfected or, if good knee-pads are being worn, it can be easier simply to crawl. Carrying tackle bags through low passages is usually easiest by towing them on the end of a short cord.

Crawling

Hands and knees crawls can be pleasant or absolute hell, depending on the type of floor. Sand or mud is luxury, smooth limestone is not unpleasant, but sharp and broken rock can be very painful. Knees are not designed for crawling and many a caver who never invested in knee-pads has been forced into early retirement. Beware of long crawls in water, such as those in Pen y Ghent Pot or Swinsto Hole, as the knees become cold and

numb – much damage can be sustained that will not be apparent until after the trip. The caver who is really suffering can try progressing backwards, sitting on their bottom.

As the roof becomes lower and lower techniques must adapt accordingly. Try to support weight on the forearms and not on the elbows, which are easily damaged. You can waste energy by dragging your body along the floor where it is not really necessary. Efficient crawling techniques can only really be perfected with practice: there is a great opportunity to experiment with different styles during a long flat-out crawl like the infamous Hensler's Crawl in Gaping Ghyll.

Squeezes

Squeezes come in an infinite variety of lengths and shapes. In their simplest letter-

On hands and knees in the phreatic passages of Carlswark Caverns. (Rob Eavis)

box form it is a case of relaxing, removing or arranging equipment into the best position, feeling for the widest part and gently easing through. If you are at your personal limit you may need to exhale and take shallow breaths as you squeeze. Remember you cannot fight your way through a squeeze: the secret is to keep your muscles relaxed and gently feel around for foot- or handholds. Downhill squeezes are usually best approached feet first, for obvious reasons.

Many squeezes are more technical than tight and can only be passed easily by adopting a certain position, lying on the right rather than the left side, for example, with one particular arm forward and the other back. Sometimes the solution only becomes apparent near the end and it is necessary to reverse back and alter position. A series of difficult squeezes, without sufficient space to turn around, can be very intimidating and demands a high level of self-confidence.

There are some classic tight caves in Yorkshire, the most notorious being Quaking Pot, Marble Sink and Stransgill Pot. The squeezes of Pippikin Pot are a bit easier and actually quite entertaining, especially on the upward journey. Although not exceptionally tight, they are awkward and the rock, as smooth as glass, gives little purchase. On Mendip, the squeeze above Primrose Pot in Eastwater Cavern is notoriously difficult.

Climbing

There is a lot of up and down underground, and climbing is an essential skill for the caver. Rock underground is free of algae and other organic deposits, making it surprisingly grippy, although liberal coatings of mud are possible. A good climbing style means staying in balance by not trying to reach too high. You should lean back slightly from the rock so that it is possible to look straight down and see both feet. The best way to improve your climbing is to practise on the surface, or train at an indoor climbing wall.

We have the advantage of usually having two walls underground, which makes chimneying techniques possible. Any distance up to two metres (if you have long limbs) can be spanned securely using, as width decreases: straddling, back and foot, elbows and knees, heel and knees, or even the soles of the boot.

Traversing

The same techniques are needed for traversing, which means horizontal climbing. It is often necessary to cross deep holes or negotiate high rifts that are impossibly tight at floor level. If there are good ledges for the feet on either side, the drop can be straddled and progress may be easy. If footholds are lacking, back and foot technique is the safest way to progress. In narrower, smooth-sided

The Vice in Ogof Daren Cilau. (Steve Sharp)

Helmet off for the upward return from the lower series of Pierre's Pot. (Andy Sparrow)

rifts the body must be held by pushing out with the arms. It is then possible to swing the legs forward and grip by pushing out with both knees. By leaning the body forward this process can be repeated.

The great Welsh system Ogof Ffynnon Ddu is well known for its traverses and the route towards the OFD 3 streamway has some spectacular situations. Another trip that requires much traversing is the route from Providence Pot to Dow Cave in Yorkshire. The so-called 'Terrible Traverse' is very entertaining for the seasoned caver, but for the less experienced it lives up to its name. Progress in many passages is often easier above floor level, and traversing often avoids much tiresome climbing up and down.

Water

There is usually water somewhere in any cave system, and it often presents dangers and difficulties. Fast-flowing water only needs to be knee-deep to be a serious hazard, and progress upstream is likely to be difficult. Many stream passages have deep pools that may be obvious in low water, but when levels rise they become invisible traps. Deep water is a considerable hazard and to swim without a wetsuit or buoyancy aid is a dangerous practice that has claimed several lives.

The resurgence at Porth yr Ogof in South Wales has been the scene of several fatalities.

The resurgence pool at Porth yr Ogof, where deep cold water has claimed several lives. (Chris Howes)

Passing a duck in Swildon's Hole.
(Carmen Smith)

The Green Canal in Dan yr Ogof. Flotation aids are essential for this long swim between vertical walls. (Brendan Marris)

An exit from this cave can be made by swimming across a deep lake, but the water is very cold and submerged tree branches and other flood debris are concealed beneath the surface. An inflatable boat is a standard piece of kit throughout most of Europe, but the British caver's enthusiasm for the wetsuit has made its use a rarity, even in caves like Porth yr Ogof where it might provide a safer option than swimming.

Ducks

A duck is a point where only a small airspace exists between the water level and the roof. In some cases, such as the 'Blasted Hole' in Simpson's Pot, the length is very short and it is simplest just to take a breath and plunge

OPPISTE: Traversing over deep rift in Upper Flood Swallet. (Rob Eavis)

BELOW: Breaking surface from Sump 1 in Swildon's Hole. (Ian Burton)

Deep joy – the mud of Ogof Cynnes.
(Brendan Marris)

through. But other ducks are much longer and require a slow and cautious approach. The 'Double Troubles' of Swildon's Hole need to be baled to give a sufficient airspace and then passed, helmet off, on the back with the nose in the roof, like a snorkel.

Sumps

A completely flooded passage is known as a sump. Most are only passable by fully equipped divers, but there are a few that are short and easy enough to dive 'free'. This is not a technique to be used casually, or at an unknown site. Unless the distance is very short indeed, a guide rope is essential and will usually be left permanently in place. It is

OPPOSITE Boats, seldom used in Britain, are the safest way to navigate deep water. (Rob Eavis)

vital to know the length, any possible difficulties, and to be absolutely sure it is the right sump and not something much longer. Whenever possible, go with someone who has free-dived the sump before.

Before entering a sump, make certain that your helmet is secure and arrange any lamp cable in a loose spiral around the front of the body to prevent catching on the roof. Give the guide rope a shake and a tug to make sure that it is not snagged (or that no one is coming in the other direction!). Exhale and inhale deeply three times before ducking under. Move slowly and carefully, pulling your body gently along on the guide rope until you surface.

In some sumps you can reach through and join hands, which can give a novice a good boost of confidence. Do not be too persuasive with a reluctant sumper – there have been several cases of novices who have gone through and then stubbornly refused to come back!

Longer sumps must be taken very seriously. Wetsuits, hoods and diving masks are essential. Swildon's Hole is well known for its varied sumps. Sump 1 is only about a metre long and is ideal for a first free-dive, but further down the streamway is the 6m long Sump 2, followed by the 11m Sump 3. Lead weights are stored by Sump 2 and the free-diver should use enough of these, evenly spaced on the belt for balance, to be just negatively buoyant. It is then possible to pass the long sumps skimming just above the mud floors and well away from protrusions on the roof.

Losing the line in a long sump could be a fatal error, but connecting yourself to it is not to be recommended as some guidelines have knots in them that may abruptly and traumatically interrupt your progress. Some sumps have small air bells, but these should be avoided due to the dangers of carbon dioxide build-up. It was bad air in an air bell that contributed to the deaths of four cavers in Langstroth Pot in 1974.

The 'Restaurant at the End of the Universe', a project camp deep in Ogof Daren Cilau.
(Rob Eavis)

Cave Camping

There are three types of underground camp:
- Bivouac camping means being prepared to rest, eat and sleep underground if it proves to be necessary, using equipment carried as part of your personal kit. The more kit you are prepared to carry the more comfortable this will be.
- Expedition camping means transportation of equipment by a team to a camp that will be maintained for a required duration and then entirely removed.
- Project camping is the establishment of a long-term underground base that is permanently maintained and regularly re-provisioned.

Camping underground allows the exploration of longer and deeper caves or permits long-duration digging and surveying at remote locations. Camps are usually associated with deep Continental systems but some recent explorations in Britain have also depended upon them. Ogof Daren Cilau in South Wales extends for several kilometres through a succession of varying passages ranging from the flat-out to the enormous. The furthest point from the entrance takes around twelve hours to reach, and there is considerable digging potential in the area, including a possible connection with the adjacent system of Agen Allwedd, just 30m distant.

The only practical way to extend a system like this is by camping, and this has been the scene of many long-duration stays, one extending to nine days. There have been other camps to aid exploration in several other of the long Welsh systems, and the technique has become almost routine in the region.

Choosing a Site

The location needs to be safe from floods, ventilated but free from draughts, dry and level. It should require no, or minimal, rearrangement or landscaping that will permanently alter the cave. A nearby water source is essential, but a small percolation dribble can be adequate for the purpose.

Equipment

Tents are seldom used: while they have the advantage of trapping a volume of air that can be warmed by carbide flames and body heat, they are difficult to pitch and bulky to carry. Specialist hammocks have been manufactured, or can be improvised. One design is completely enclosed and incorporates a suspended candle holder that warms the interior, making a sleeping bag unnecessary.

Usual practice is to lay out a plastic sheet to provide a clean area, and then to use either an air bed or a closed-cell sleeping mat. The latter, though bulky, weighs nothing and cannot be punctured or deflated. The sleeping bag should be of synthetic fibre rather than down, which is too prone to absorb moisture. A dry spot can often be found under an overhanging wall, but if drips cannot be avoided a second plastic sheet may need to be stretched out over the sleeping area.

Sleeping bags and dry clothing can be carried in sealed plastic drums, or even tied up in bin-bags. A waterproof bag, as used by canoeists, works well. Spare, dry clothing can be carried in the same way, and used both for sitting around at camp and as pyjamas

when sleeping. A balaclava will reduce heat loss from the head and keep the sleeper much warmer.

If the camp exceeds a couple of days, dry clothes may run short. The only way to dry clothing in the humid conditions is to wear it and let body heat do the job, which may mean sleeping in damp clothes. Ideally the caver should have several dry long-sleeved thermal vests that can be worn under the damp clothing, thus greatly increasing comfort.

I have it on good authority from a veteran of many underground camps that the most efficient cooking stove is the good old primus. For those who prefer a simpler option, Butane gas stoves using self-sealing cartridges are probably the best choice. It is possible to cook using a carbide flame, though this is probably best saved for emergencies. All water should be sterilized.

Hygiene

The longer the camp is in use the more important hygiene becomes. If the floor is

Dry clothes and a fry-up prepare a team physically and psychologically for another day of exploration deep underground. (Steve Sharp)

The environment of Lechuguilla Cave in New Mexico, USA, is so fragile that all contaminants, even crumbs of food, must be collected and removed from the cave. For this reason the camping area is confined within marker tapes.

sand or mud a latrine can be dug, but otherwise a toilet will need to be contrived. Urine can be collected in an empty bottle and emptied where it will have least effect, either into an excavated hole or into flowing water. A small plastic drum half-filled with cat litter can be used for solids, and then carried out of the cave.

Camping Ethics

Where camping serves a purpose and contributes to exploration it is a justifiable technique, but it can have a very negative effect on the cave environment, and it should not be undertaken simply for novelty value. The old problem of items being much easier to carry in than out is starkly illustrated in some underground campsites. The Hall of the Thirteen in the Gouffre Berger is the most commonly used underground camp in Europe. This beautiful chamber is despoiled by mounds of rubbish, heaps of spent carbide and the pervading odour of human waste. The fact that this cave can only be explored by experienced cavers says little for conservation standards. It is not surprising that the caving community feels uneasy about sponsored cave-ins and charity cave camps.

CHAPTER 6

Vertical Techniques

Virtually all caves have some kind of vertical hazard, ranging from steep slopes to gaping chasms, and techniques vary as required, from simple handlines to the use of complex single rope climbing systems. Most caving ropework is based on a few simple knots and easily learnt techniques. These are the foundation of safe vertical caving and they should be the starting point for every caver. Most falling accidents could be prevented by the correct use of these simple techniques, which can be learnt in just a few hours.

Ropes

There are two types of rope construction, the traditional hawserlaid and the more modern kernmantle. You are more likely to find hawserlaid rope in a builders' merchant than in a climbing or caving shop. It is not suitable for our purposes for a couple of reasons: firstly it has a tendency to unwind, which causes an unpleasant spinning during suspension, and secondly it cannot be used safely with some abseil or belay devices.

Our ideal rope is going to be of the kernmantle construction, in which nylon fibres are contained within a plaited sheath. If the fibres within the sheath are tightly spiralled, like a coiled spring, the rope is described as 'dynamic', but if they are more loosely spiralled the rope is described as 'static'. It is very important to understand the difference

between these types. Dynamic rope is designed for climbers and is very stretchy and shock absorbent, while static rope is intended for abseiling, prusiking, top-rope lifelining or any situation where there is no danger of sudden and extreme shock loading.

There is no simple way to tell the difference between the two rope types. Generally, static ropes have a more tightly woven and inflexible sheath and are usually manufactured with a simple colour scheme, while dynamic ropes have a slightly softer feel and are often produced in lurid colours. The difference between the ropes only really becomes apparent when they are used to

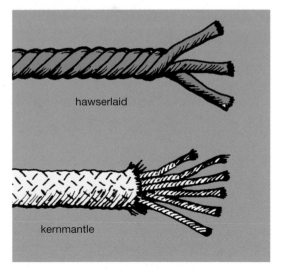

hawserlaid

kernmantle

Rope types.

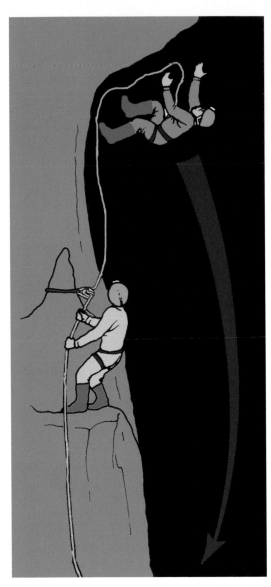

Length of fall 10m, length of rope 5m. The fall factor is 2 or 200 per cent.

the illustration shown here, the lead climber will fall twice the length of the rope before being held by the lifeline system. The further he falls the more he accelerates, the greater his momentum and the harder he lands! Rope absorbs the energy of a fall by stretching. This converts any sudden impact loading to an event with a longer duration but with much lower forces generated – exactly the same principle as a bungee jump.

The result of this effect means that distances fallen are only meaningful in relation to the length of rope available to hold the fall. A 10m fall on to a 30m length of rope produces low shock loadings, while a 10m fall on to 5m of rope, as illustrated, creates the highest possible shock loading. To estimate the severity of a fall we need to apply the principle of fall factors.

The severity of a fall can be calculated by dividing the length of the rope by the length of the fall to give a value between 0 and 2. This is known as the fall factor. It is easier, though, to think in percentages and consider the fall as a percentage of the rope. Examples of this principle are also illustrated. Dynamic rope is designed to stretch and absorb the energy from the high fall factor often experienced by climbers. Not only will the rope not fail, but also the climber should be able to walk away uninjured.

Imagine a caving situation: a caver is being lifelined up a ladder pitch but climbs too fast for the lifeliner to take in the rope, and then slips and falls. He falls 3m (having created only a 1.5 metre loop of slack) on to 10m of rope and generates a fall factor of 0.3 or 30 per cent. A fall of even this relatively low fall factor could cause serious injury if static rope is being used.

Imagine another situation: to continue, the cavers, who are using static rope, must follow a rising traverse, over a shaft and into a higher level passage. If the caver slips he will fall nearly 5m on to 2.5m of rope – a

support body weight. The static rope will stretch to about 3 per cent of its length under an 80kg load, while the dynamic is more elastic and will stretch to around 6 per cent.

To appreciate the significance of this difference we need to understand the implications of shock loading in different situations. In

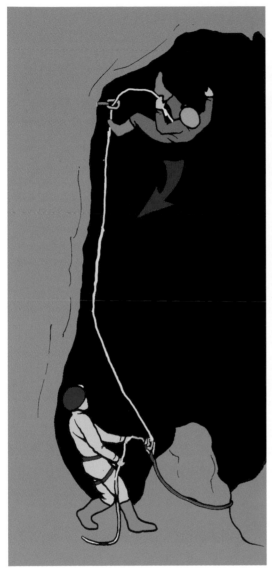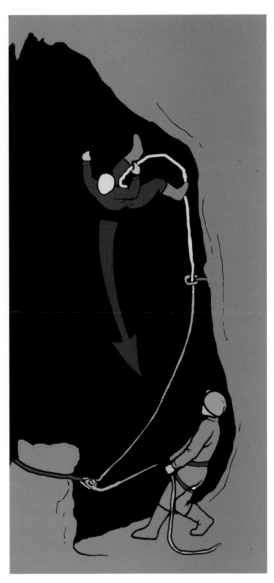

left-hand: Length of fall 5m, length of rope 15m. The fall factor is 0.33 or 33 per cent.
right-hand: Length of fall 10m, length of rope 10m. The fall factor is 1 or 100 per cent.

fatally high fall factor. How is the danger to be reduced? Assuming the rope is long enough, the lifeliner moves to point A, putting an extra 15m of rope into the system. This reduces the fall factor to 0.15 or 15 per cent, and greatly reduces the danger.

Static ropes are now routinely used as multi-purpose caving ropes. They are safe and suitable, but only if high shock loads are avoided. The simplest way to achieve this is by effective and attentive lifelining, coupled with an understanding of the principle of fall factors.

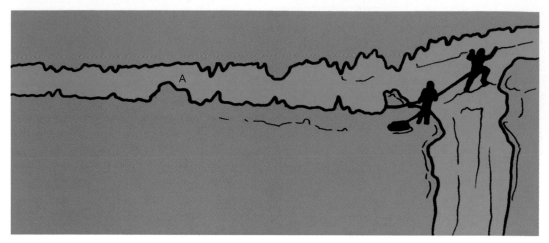

The fall factor in this example can be reduced by lifelining from point A.

Diameter

Caving ropes range in diameter from 8mm to 11mm. As the diameter decreases, the sheath of the rope becomes progressively more significant as a load-bearing component. The general wear and slight abrasion on the sheath of an 11mm rope may not be significant, but with 9mm or 8mm rope it can indicate a significant loss of strength.

Thinner ropes are light and relatively cheap. Experienced cavers find them useful because they allow smaller teams to rig deep caves, but they do require considerable skill to use safely, and their shorter working life can negate any initial monetary saving. They are not intended for use with some belay devices or descenders.

8mm Static. This is intended for expert use only, it is very easily damaged by abrasion and has a short working life. It is intended for ultra-lightweight SRT trips into deep caves. Also used for making rope slings, deviation cords and SRT footloops.

9mm Static. This is a more practical option, and is more commonly used. It is still very liable to abrasion damage and is not recommended for general use, although it is handy as an emergency rope, a 15m length occupying very little space.

9mm Dynamic. High stretch makes this unsuitable for SRT work. It is very prone to abrasion damage and not suitable for caving use. 9mm dynamic ropes are classed as half ropes and are intended only for use as part of the double rope system used by climbers. Sometimes used for cowstails, but not really recommended for this purpose.

10–10.5mm Static. This is now the standard for most caving. It gives the best compromise in terms of, strength, stretch and working life.

10mm Dynamic. This can be used for very short SRT pitches and general lifeline use, but its main function is for cowstails.

11mm Static. A bulky but hard-wearing rope that will tolerate some abrasion. It has the lowest stretch of the available ropes and this makes the effects of shock loading all the more serious.

11mm Dynamic. Suitable for use on very short SRT pitches, for general lifelining and cowstails, if you like them chunky.

Rope Care

Look after your rope, and it will look after you. Keep the rope as clean as possible underground by always carrying it in a tackle bag. Wash it after use and remove all particles of grit by pulling it between two scrubbing brushes, or by using a rope washer designed for the purpose.

Ropes should be protected from direct sunlight and stored in the dark as prolonged ultraviolet radiation is very damaging. Beware also of chemical contamination, especially from caving light batteries. Lamp and rope storage should be completely separate to minimize this danger.

A simple rope-washer. The brush inserts into the tube, which is held underwater by using a foot through the loop.

Ropes should always be carried in tackle bags to protect them from mud and grit. (Brendan Marris)

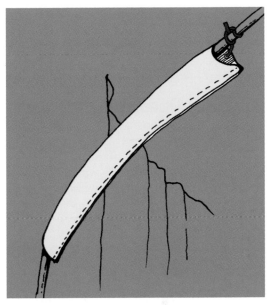

A rope protector is a wraparound sheath positioned over a sharp edge.

Shrinkage

New ropes can shrink by up to 15 per cent following their first immersions in water. One beneficial effect of shrinking is to tighten the sheath on to the core, preventing one slipping in relation to the other. It is advisable to soak ropes for at least 24 hours before use, and before cutting and marking specific lengths.

Marking

Ropes should be marked clearly with their length, date when purchased and any reference number. The labelling can be done on white insulation tape with a waterproof marker, and then sealed with clear heat-shrink plastic. Even after the rope has been soaked some further shrinkage is likely, so it is advisable to understate the length by 10 per cent.

Testing Ropes

The breaking load quoted for new ropes is tested on a strain gauge in a laboratory and has little to do with real working conditions. We cannot use a rope without knotting it, and most knots effectively halve the strength of the rope. We are unlikely to keep our ropes dry, and tests have shown that wetting has a chemical effect that reduces rope strength. A brand new rope will not be new for long and after a relatively small amount of use underground, the strength of a rope (as measured by drop tests) can be reduced by up to 75 per cent.

The only effective way to test a rope is by a destructive drop test using a 2m sample, usually with a figure 8 knot in each end. Many clubs have their own drop test rigs, which are used for systematic routine testing of ropes over a certain age (usually two–three years for 10mm). The test uses an 80kg weight for a fall factor 1 (or 100 per cent) drop. Generally, the sample is expected to survive at least two drops before the rope is declared safe.

Means of Attachment

Belts

Back in the early days there was only one way to attach to a lifeline and that was by a direct tie around the caver's body. While this simple technique undoubtedly saved many lives, it was crude and limited. An exhausted ladder climber could not comfortably rest or be assisted by hauling when suspension in this way caused considerable pain and even difficulty breathing. A simple innovation arrived in the 1970s with the load-bearing or 'belay' belt – a 5cm wide length of webbing, industrially sewn, with a secure buckle. Using a belt the caver could simply clip in to the lifeline with a karabiner (avoiding the hazard of wrongly tied knots) and experience much less discomfort when resting. Thus it was that the belt became standard personal equipment for most cavers.

As time moved on, and with the advent of SRT, the greater benefits of the harness became apparent. For cavers in predominantly vertical caving areas the harness became standard personal kit, but in regions with fewer, shorter and isolated pitches the simple belt was still favoured. Personal Protective Equipment (PPE) legislation, introduced in the 1990s, did not recognize a belt as being adequate for human suspension, as a result of which caving belts were not CE marked and could no longer be recommended by the manufacturer for this purpose.

In reality very little has changed in the use of belts by cavers. Manufacturers no longer sell 'load-bearing' belts but they do supply 'battery belts', which are, coincidentally, a 5cm wide length of webbing, industrially sewn, with a secure buckle. It is not really surprising that cavers are reluctant to be parted from their belts, for what are the alternatives? Either it becomes standard practice to carry or wear a harness for every caving trip or we return to tying ropes around our waists.

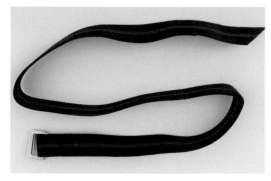

A typical caver's belt.

Harnesses

These are an essential item for cavers using single rope techniques (SRT) and are a godsend for any tired caver struggling up a ladder pitch. The sit harness will support a suspended caver quite comfortably for several minutes, or longer if a fully padded climbing version is used. Be aware that climbing harnesses are not suitable for SRT use. Caving harnesses are always secured at the front by either a semi-circle maillon rapide or a purpose-designed semicircle karabiner. Most caving harnesses have a lower centre of gravity than a climbing harness, which can cause a top-heavy person to invert. This danger can be reduced by using a large diameter karabiner to clip into the semicircle maillon and around a securely buckled load-bearing belt.

There are a couple of ways of improvising a harness using a sling that work very well. The 'nappy seat' made from a standard-length tape sling fits most cavers, but with a shorter sling (or larger bottom) try the figure 8 method. This only works if you are using a securely buckled load-bearing belt.

Cowstails

Cowstails are tied from 2.5m of dynamic rope using three knots to create a long and a short safety connection. Figure 8 knots can be used or you may prefer the self-tightening 'barrel' knot. For ladder and lifeline use the lengths are not critical, but neither should exceed about a metre. For SRT the lengths are more precise (*see* Chapter 7).

Caver's cowstails.

Traverse line and cowstails in use on a ledge traverse. (Brendan Marris)

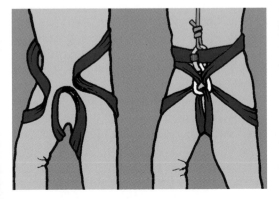

An improvised harness connected to a belt that raises the centre of gravity.

HMS or Pear-shaped
Karbiner
Uses: general but
intended for Italian
nitch.

Correct loading of
karabiners is important;
they are designed to
bear two-way loading
along their major axis.

Oval
Karabiner
Uses: general
but required
for the correct
use of Bend
bolt hangers
and for the
assembly of
pulley-jammer
systems.

D-Shaped
Karabiner
Uses: rigging and
general use.

Maillon Rapide
An alternative to the
karabiner; useful in
many rigging
situations. Ideal for
multi-directional
loading.

Types and uses of karabiners.

Karabiners

These allow quick, safe and easy connections to be made. They are available in various shapes and sizes intended to perform particular tasks. All karabiners ('krabs') are intended to be loaded along their long axis and never across the gate, which is very much weaker.

Maillon Rapides

Maillons are slower to use than karabiners but they are cheaper, smaller and have the advantage of allowing multi-directional loading. They are made in various shapes and sizes, the smallest normally used for rigging being the 7mm long opening. They must be fully closed for safety, which may require the use of a spanner. They are most useful for rigging in positions where they only need to be screwed and unscrewed once during the trip.

Slings

These are loops of webbing or rope often used in rigging. Sewn webbing slings with a circumference of 3m are most commonly used, though in practice these have little advantage over rope slings tied from 8 or 9mm rope using a double fisherman's knot. Tape slings are usually sewn and should only be tied using the correct tape knot (many conventional knots are unsafe if they are tied in tape).

Anchors

The rope can be no stronger than the belay point to which it is tied. A jerky abseil descent can generate loadings of up to five times body weight, which means a heavy caver could be putting half a tonne on the anchor. Belay selection is a vital skill that many inexperienced cavers seem to lack. The brittle and often fractured nature of limestone demands much caution. You can try tapping the anchor with a karabiner or spanner and compare the sound with the surrounding area – does it 'ring' or does it sound dull and hollow? Don't be afraid to test anchors physically by tugging or kicking, to check that they are as solid as they appear, and that there is no suggestion of any movement.

A single massive anchor is fine, but sometimes the choice is between several lesser options and distributing the load evenly between them will be necessary. The multiple bowline is one of the simplest ways to achieve this and, with a little practice, can be tied and adjusted very quickly. An alternative – useful for saving rope – is to use a sling and karabiner around one or more anchors. Remember that two dodgy anchors do not equal one safe one!

When using two or more anchors it is important to share the load evenly between them. An angle of 90 degrees between the two arms of a Y hang, as this type of arrangement is often called, will load each point with 70 per cent of the weight suspended on the rope. If the angle is increased to 120 degrees the load at each anchor increases to 100 per cent, and at 150 degrees the load at each anchor reaches 200 per cent. A badly rigged shared anchor can actually increase the loadings on the anchor points rather than reducing them.

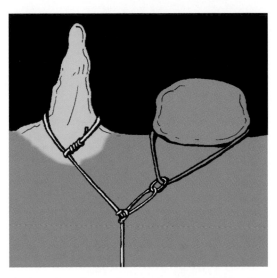

A shared anchor – the ideal angle between the two arms is 90 degrees.

Bolt Anchors

Fixed anchors are common in the more popular caves. The term 'bolt' is applied to several types of artificial anchor point. There are expansion bolts (rawlbolts), 12mm self-drilling anchors or 'spits', and the resin-bonded 'P' anchor or 'Eco-hanger'. Any correctly installed and maintained anchor is, initially, more than strong enough for the job. Eventually, though, metal corrodes, limestone fractures, resin cracks, and any anchor can fail, hence the simple and essential rule: Never trust your life to a single bolt of any kind! In view of this principle, all bolts, even the massively strong resin anchors, are routinely installed in pairs.

There has been an inevitable trend towards resin anchors in the more popular caves, but the ability to use 12mm spits correctly still remains a useful skill. Resin

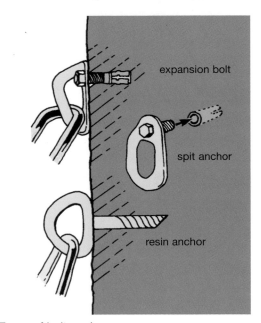

Types of bolt anchor.

anchors are easily located but spits can be virtually invisible and much experience is required to anticipate their position. The threaded sheath is embedded flush to the wall and usually left without the essential

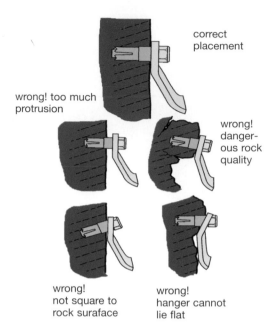

correct
placement

wrong! too much
protrusion

wrong!
danger-
ous rock
quality

wrong!
not square to
rock suraface

wrong!
hanger cannot
lie flat

Correct and incorrect bolt placements.

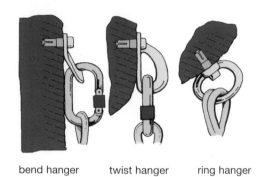

bend hanger   twist hanger   ring hanger

Types of hanger.

Stainless steel, resin-fixed 'P' anchors are now installed in many of Britain's most popular vertical systems. (Tom Philips)

'hanger' – an attachment ring or plate for the rope that must be screwed into place (you will need a 13mm spanner). Of the various hangers available the ring and twist types are the most versatile. Check the anchor before use; it should be flush and square to the rock, clear of mud and grit, and at least 15 cm clear of obvious cracks or weaknesses. Hangers should be kept clean and the bolts lubricated. Do not overtighten the bolt: finger-tight plus half a turn is sufficient.

Do not attempt to install any new anchor in a known cave without consultation with relevant conservation and access bodies. There are already far too many poorly installed anchors, in useless positions, placed by cavers who lack the basic skill to find and use those already existing.

Knots

The Three Basic Knots

There are three simple knots that every caver should know: the bowline, figure 8 and

Italian hitch. If you are unfamiliar with these, get yourself a short length of rope or cord and practise until they become fluent.

1. Bowline

Bowline.

This is quick and easy to tie, to adjust and to untie. It is commonly used for anchoring around thread belays (and for the traditional direct tie around the body). The tail of the rope should be secured as shown to provide a safety locking knot. Alternatively, the tail can be used to tie another bowline around a second anchor to provide higher safety. This technique is known as multiple bowline. A double bowline is the same knot tied in a double rope. This is more difficult to adjust for shared loading than the multiple bowline, but is useful when one anchor point is much nearer the pitch-head than the other.

2. Figure 8

This is commonly used to provide a clipping-in loop in either the end or middle of the rope. When used as an end knot a stopper knot is advisable.

3. Italian Hitch

A very useful knot that acts as a friction hitch. It can be used for abseiling or self-protecting short climbs but its main function is lifelining. The knot should be used with a large diameter karabiner (D-shaped steel or

Figure 8.

Italian hitch.

HMS pear-shaped alloy), which allows it to flip over and change direction. The Italian hitch is easily locked-off by a couple of half hitches.

Other Useful Knots

Butterfly

A very quick and easy-to-tie knot that is used for general rigging. It is similar, but not iden-

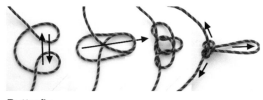

Butterfly.

tical, to the Alpine butterfly, a knot often used in climbing. It is used ideally either for a three-way loading or for a mid-rope tie-in (a 'middleman's knot').

Bowline on the Bight

The best knot for tying a shared belay or Y hang, especially between bolt anchors. It can also be tied by rethreading and used for shared loading around thread anchors.

Bowline on the bight.

Double Fisherman's

The only recommended knot for joining ropes, this is particularly safe for connecting ropes of different diameter. Half a double fisherman's can be used as an excellent stopper knot in conjunction with the bowline or figure 8. Half a double fisherman's can also be used to create a noose sometimes known as a 'barrel knot', which is used to secure karabiners to cowstails.

Double fisherman's.

Tape Knot

Many conventional rope knots are unsafe when tied in webbing, and this is the only recommended method for joining tape. Generous tails of at least 10cm should be left, and the knot pre-tightened by hand before use. It is a good policy only to use pre-sewn tape slings, which are stronger and safer.

Tape knot.

Capuchin or Stopper Knot

This large ball-shaped knot makes an ideal stopper, as it is sufficiently large (tied in 10mm rope) not to pass through a figure 8 descender, or even an HMS karabiner.

Capuchin or stopper knot.

Prusik Knot

This is tied using a 3–6mm cord. It is simple but ingenious as it can be slid up and down the rope, but (usually) binds tight when loaded.

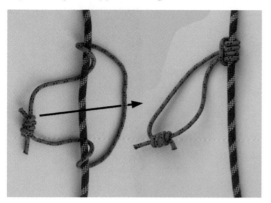

Prusik knot.

Klemheist

This is an alternative to the Prusik knot that uses tape and provides an alternative if cord is not available. It is much stronger than the conventional Prusik knot.

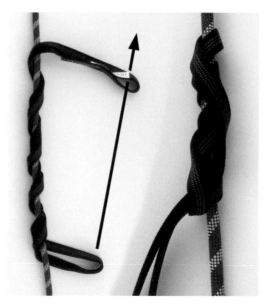

Klemheist.

Clove Hitch

A quick and simple way to tie into a karabiner using minimal rope, appropriate for mid-rope, rather than as an end-knot. Not for general usage, but occasionally very handy.

Clove hitch.

Overhand Knot

A loosely tied overhand knot is as strong as a figure 8, but after loading it is much weaker, and very difficult to untie. It is useful for tying a quick loop which can be clipped into, briefly, for safety.

Overhand knot.

Warning!

There is often much loose rock precariously balanced at the pitch-head, or on ledges partway down. It is very important not to stand around in an exposed position at the pitch-bottom, but to find a sheltered spot, which is protected by an overhang. Likewise, great care is required at the pitch-head not to dislodge anything on to those below. In some caves, such as Rhino Rift in the Mendip Hills or Hangman's Hole in Yorkshire, the danger is acute. Such caves are best suited to small, experienced parties who can recognize and minimize the hazards.

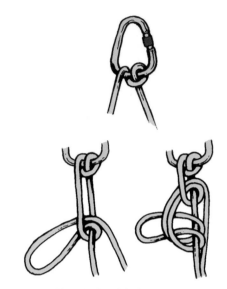

Locking-off an Italian hitch.

Lifeline Systems

We need a system that transfers the loading directly to the anchors and functions regardless of the relative weight of the climber and the lifeliner. It should lock off, allowing the operator to go hands free, and permit instant, controlled lowering. It must also be a system that can be operated easily and quickly enough to keep up with a fast ladder climber. Other useful refinements include auto-locking and low friction, which allows the use of counterbalance systems to assist the climber. The system used should be designed to function at head height.

Italian Hitch

This is the simplest and one of the most effective ways to lifeline. All that is required is a karabiner with a sufficiently large diameter to allow the knot easily to flip over and change direction. Pear-shaped, purpose-made HMS karabiners are available. These are almost exclusively alloy and do not survive well when used with muddy or gritty ropes. They often have 'quick-lock' gates that do not function well when infiltrated by cave grit. A large D-shaped steel karabiner works reasonably well for Italian hitch use.

Operating the system is relatively simple but does require some practice. The Italian hitch works on a self-tightening, but not self-locking principle. It requires only a small amount of tension on the 'dead' rope side to hold a considerable load. Safe operation depends on the correct handling of the dead rope. When lifelining downwards, control is entirely on the 'dead' rope side, but when lifelining upwards one hand is needed to feed in the 'live' rope. The rope can be quickly and safely locked off, allowing the operator to go hands free.

This system does not convert to hauling as quickly or effectively as some more technical lifeline methods, but there is a fairly simple way to assist tired climbers on the upper section of a pitch. The hitch should be locked off and the emerging dead rope lowered in a loop to the climber. A karabiner clipped into this loop acts as ballast and is used by the climber to connect to their harness, or is

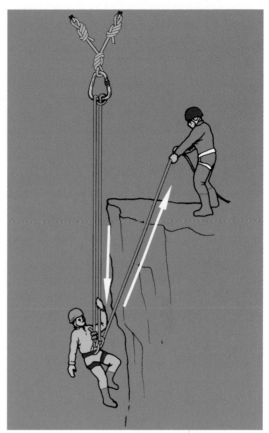

Converting an Italian hitch to a hauling system (1): the hitch is securely locked-off before lowering the karabiner.

Converting an Italian hitch to a hauling system (2): the system will work most effectively if the climber can pull the rope through the hitch from below.

clipped into the karabiner connecting them to the lifeline.

The hitch is unlocked and held securely in the usual way, while the slack in the loop of dead rope is taken in. The lifeliner can then haul on the new dead end, and the effect is a 2:1 self-locking system. The safety of this system is dependent on the climber connecting safely to the lowered karabiner, and if this cannot be visually confirmed from above the technique should not be used. Practice is essential before use.

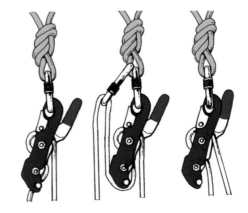

Stop descender used for lifelining. The different threadings provide varying friction.

Stop Descender

The stop descender, used by many cavers for abseiling, also functions very effectively as a lifeline system. Users of the stop will be aware that its auto-lock mechanism does not function well on some new, dry or thin ropes, and it is important that these are not used when lifelining with this device. The stop should be rigged at head height, out of contact with the walls. The correct threading of the descender is far from obvious and is easily done incorrectly. The ideal threading system for lifelining differs from that normally used for abseiling by reducing the friction and allowing easier operation. This can still provide full friction by use of a braking karabiner, but it allows an instant change-over to a lower friction system, which is more effective for lifelining upwards.

The lifeliner should use both hands to operate the system, both up and down. When used for descent, one hand must depress the handle while the other controls the dead rope. It is then necessary to react to any sudden increase of speed, indicating a fall, by releasing the handle. The operator should then be prepared to grip the dead rope firmly with the other hand. On the upward climb the braking karabiner is removed and both hands are used to take the rope in. In the event of a fall, lock-off should be automatic. Assuming the stop is rigged at head height, it is possible for the lifeliner to assist the climber by leaning back and applying body weight to the rope. By connecting to the dead rope using an Italian hitch, prusik or jammer, counterbalance can be used to assist or lift the climber. Conversion to a 'Z-rig' is easy with this system.

Grigri

This device is similar in principle to the stop descender but is specifically designed for

The Grigri.

lifelining. It has the advantage of locking-off on any rope including new and dry ones (in the author's experience to date), but is not recommended for ropes of less than 10mm. It is very simple to thread and to use (and also quite easy to thread incorrectly). It does have the disadvantage of needing to be completely unclipped each time it is rethreaded, which presents a danger of dropping it down a pitch. A very useful device, but seldom seen underground.

Mini-Traxion and Pulley Jammer

A pulley jammer is a device or arrangement that combines these two elements together to create a low-friction self-locking system. A mini-traxion is the simple way to achieve this, but you can put one together using separate items. For this you will need a fixed cheek pulley, an oval karabiner and a Petzl basic or handled jammer. The system for fixing them together can be remembered as POOO (Pulley to Oval karabiner, Oval karabiner to Oval hole in jammer). It is vital to check that the rope is running through the system in the right direction.

Head-height rigging is essential or lowering off may be impossible. The system is for lifelining upwards only and has the advantage of being almost frictionless. The opera-

The mini-traxion or a pulley jammer assembled from separate components both fulfil the same functions. A very useful device in skilled hands, but potentially lethal for the untrained!

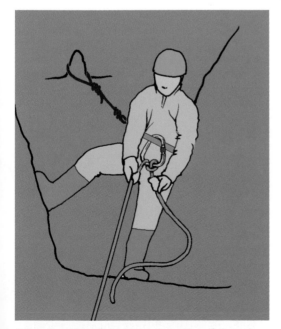

Lifelining directly is the only option if there is no absolutely safe anchor. Any small attachment point that is available can be used to help secure the caver in position.

tor can assist the climber by leaning back and applying body weight to the rope. By connecting to the dead rope using an Italian hitch or jammer, counter-balance can be used to assist, or even lift the climber.

The danger with the system is the difficulty of releasing and lowering in an emergency (for example, an exhausted climber suspended under a waterfall). To achieve this, the operator must connect to the dead rope with a locked-off descender or Italian hitch, counterbalance, release the jammer and lower off. If the climber is heavy, a second jammer or prusik around the live rope will be needed to pull against. This system must be surface-practised, and should only ever be used when both climber and lifeliner are using harnesses, never belts.

Body Belay

This is the traditional and outdated system that fails to meet any of the requirements of a modern lifeline system. It has only one advantage in that it requires no karabiners and is included here as an emergency technique. The lifeliner should be tied on and well braced. The live rope should pass through one hand, go around the lifeliner's back, have one turn around the wrist and then be held in the second hand. The feeding action illustrated should ensure safe

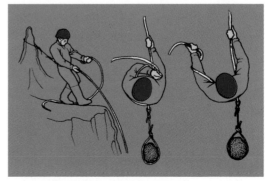

Body belay system – best avoided but can be used in an emergency if no other equipment is available.

operation. If the rope is wrapped around the wrong wrist, a sudden load could easily cause serious injury.

Z-Rig

This is an easy way to convert the lifeline system into a 3:1 lifting system and is invaluable for assisting tired climbers. Connect a jammer or prusik knot to the live rope 1 or 2m below the lifeline system, then connect the emerging dead rope to this with a kara-

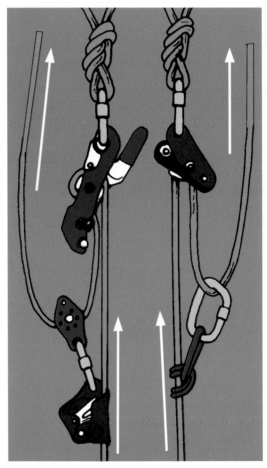

Z-rig hauling systems based on the stop descender and grigri. The advantage of using these devices is the ease of release and lower.

biner or, better, a pulley. Pull up on the dead rope and you have the lifting system. It is necessary to run the jammer or prusik back down the live rope after each pull, but if a tackle bag or spare gear is clipped to the jammer (it doesn't work with a prusik) it will run down under its own weight. Z-rigs can be used with all of the recommended systems with varying degrees of effectiveness. Because the Italian hitch does not self-lock it needs to be held manually between each lift. You will only achieve a 2:1 advantage using an Italian hitch because of the friction created within the knot. Stop descenders and grigris are more efficient and have the advantage of easy release and lower. A pulley jammer works most effectively but release and lower is much more difficult. The system should only be used by cavers familiar with the technique of release and lower, and only from anchors positioned to facilitate this (head height directly above the pitch). Any hauling system must connect to a harness and never a belt.

Other Systems

There are almost too many methods, systems and gadgets to choose from. You cannot buy or borrow technical equipment and expect to use it safely without training or, at the very least, carefully reading the instructions. Somebody may be a great enthusiast for a particular system and recommend it keenly, but have they considered all its implications? There have been several accidents caused by using unsuitable lifeline methods. For example, a caver fell and was injured because a belay plate (a system favoured by climbers) was secured at head height rather than waist height. Another injury occurred when a shunt (intended as an abseil safety brake) was used to belay. This device locked off automatically but could not be released to lower the exhausted climber down. When eventually the release handle was forced open, the climber dropped out of control. I am particularly sad to report a recent fatality in which the inability of a group to release a loaded mini-traxion seems to have been a component.

Climbs, Slopes and Chimneys

Many drops underground are easily free-climbed while others are completely vertical. Between these two extremes are a multitude of slopes, chimneys and pitches where techniques will vary according to the judgement and abilities of individual cavers. For these minor climbs there are three main options, each with its own merits and demerits.

1. Simple Handline

This is just a rope (slightly longer than the drop) anchored at the top and usually used in a hand-over-hand method to assist in ascent and descent. An Italian hitch (using a karabiner on the caver's belt or harness) can be used to make the descent more controlled. Advantages: simple system, minimum kit, quick. Disadvantages: no security, difficult for tired cavers.

2. Assisted Handline.

Similar to the handline but with the rope running through a karabiner fixed to the climber. This allows a controlled descent and gives much assistance for a tired climber on the ascent. Ideal for short, steep (but not vertical) slopes. The lifeliner can clip on for safety using a long butterfly loop tied close to the anchor. Advantages: very safe, very helpful for tired cavers. Disadvantages: needs more rope, liable to friction if the ropes are crossed or twisted.

Simple handline.

Assisted handline.

3. Italian hitch belay

This is the basic caving lifeline system that all cavers are advised to adopt for standard use. Ideally, it should be rigged with the Italian hitch at head height, but any position that leaves the karabiner clear of the walls or floor will work. The system of operation is described in detail above. Advantages: complete security. Disadvantages: does not easily convert to haul, no handline to assist in climb.

Scrambling up a waterfall in Agen Allwedd using a simple handline. (Steve Sharp)

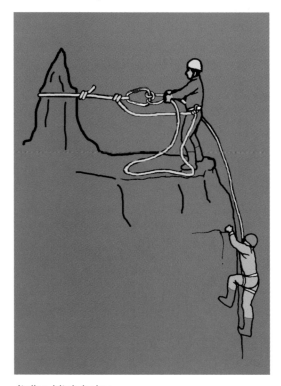

Italian hitch belay.

Climbing Calls and Whistle Signals

Poor acoustics, powerful waterfalls and the possibility of lifeliner and climber being out of sight from each other, make good communication essential. Many accidents have been caused by a failure to learn and use the correct climbing calls. Every caver must know and routinely use the standard system as listed below:

CLIMBER: (*having clipped into lifeline*)
'Take in' (*when lifeline pulls tight*) 'That's me'
LIFELINER:
'Climb when you're ready'
CLIMBER:
'Climbing'
(*but waits before climbing*)
LIFELINER:
'OK' (*climber starts to climb*)

Other useful signals include:
If the rope is not taken in fast enough:
'Take in'
 If slack is required:
'Slack'
If anything is dislodged or falls down the pitch:
'Below'
When returning the rope to the pitch bottom:
'Rope below'

Whistle signals are seldom used but may be useful on noisy wet pitches or shafts with poor acoustics. Remember SUD:

One blast = Stop Two blasts = Up (take rope in)
Three blasts = Down (let rope out)

Ladders

The flexible caving or 'electron' ladder is often used for short pitches. A 10m length can be rolled into a compact bundle weighing only a couple of kilograms. Extra lengths of ladder can easily be added by using the C-links, and the system extended as required. Ladder climbing requires a good technique to keep the body upright and reduce the strain on the arms. The rungs are gripped from behind with the hands held at face level, one foot is inserted from the front and one, heel first, from the back.

Even with an ideal climbing style it is very hard work. In the days before SRT, ladders were used on shafts of over 200m, but, realistically, the ordinary mortal finds 30m more than enough. Novices or cavers with a poor power-to-weight ratio can find even 5m of ladder a major obstacle.

Falling from ladders is common (almost routine for beginners) and the lifeline is therefore a vital component of the ladder system. The single biggest cause of death and injury in the history of British caving has

50m ladder climb in Juniper Gulf. It is now very unusual to use ladders on a pitch of this length. (Clive Westlake)

been falling, unlifelined, from ladder pitches of less than 10m. The victim has often been inexperienced and in the company of cavers who have omitted a basic safety precaution for the sake of carrying a short rope and taking a few extra minutes.

Ladders are subject to corrosion and consequently to sudden failure. This is also true of the other components used in the ladder system – wire belays and spreaders that are sometimes used for anchorage. It is important that these items are not used as links in the lifeline system as they have no shock absorbency and can fail under relatively low loads. There are good arguments for dispensing with these items completely; the spreader can be replaced with two maillon rapides and the wire belay can be substituted by a short length of rope or by using tape slings.

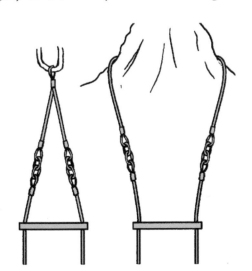

Ladders can be secured using either a 'spreader' or a steel tether.

Rigging a Ladder Pitch

There are two different techniques for rigging a ladder and lifeline system. The method used will depend on the depth of the pitch, whether it is wet or dry, the proficiency of the party and the available equipment.

1. Rigging Short, Dry Pitches

The first objective is to find a sound anchor (or anchors) suitable for lifelining. Since the group is not likely to be equipped with cowstails, this will need to be a safe distance back from the edge. Having arranged the belay, it can be used to protect anyone approaching the pitch-head, such as the ladder rigger. The ladder often requires a separate anchor to give a good position. The ideal ladder position should make getting on and off fairly easy and it should hang away from falling water or loose rock. Ladders are most easily climbed when they hang just away from a wall.

A simple rig from natural anchors suitable for a short dry pitch and a technical rig that avoids water and can be easily converted to hauling.

A single bolt is often used for anchoring the ladder. This is acceptable because the ladder is not part of the safety system and assumes that all the climbers will be safeguarded by a lifeline. Unroll the ladder from the pitch-head rather than letting it unroll itself. This prevents the ends from tangling, or falling back between the rungs, which causes damaging permanent kinks in the wires.

A double lifeline system is required to protect the last person down and first back up. The rope is run through a pulley (a karabiner can be used on shorter pitches) that must be anchored out over the pitch to allow free rope movement. A general rule is that you should be able to see the pulley from the bottom, otherwise the system is likely to jam. The last but one person down can rig the rope through the pulley before descending while being lifelined from above. A lifeline must then be arranged from the pitch-bottom, which can be from natural or bolt anchors, but it is more common practice to lifeline directly from the belt or harness. The lifeliner can connect to another person to act as ballast if they are significantly lighter than the climber. Standard climbing calls must be used at this point between the lifeliner and climber.

On the ascent this process is reversed. The first man up leaves the rope rigged through the pulley while he re-rigs the lifeline belay. The second man is lifelined from above and he removes the pulley, while protected by the lifeline. These procedures ensure that every member of the party is protected on, approaching, and above the pitch. Standard climbing calls must be used.

Rigging a pitch as described is usually easiest if a short rigging rope is used. A 4–5m length can be used to link anchors together and, often, to position and back-up the double lifeline pulley. A single short rope is usually more useful than any number of slings and wire belays.

2. Technical Rigging

The technical rig usually relies on bolt anchors to place the ladder in a very precise position. Harnesses and cowstails will be essential for every member of the party. A traverse may be necessary to reach the hang point (especially if the pitch is wet) and rigging must then begin with a traverse line. Traverse lines, or sections of traverse lines, can be rigged for protection or aid.

Protection traverse lines fall into the two categories of shock loadable and non-shock loadable. A line that prevents group members from venturing near the edge is non-shock loadable, and can be safely used by several people at once.

Most traverse lines will not prevent a slip or fall and shock loading is possible. If the line is rigged at head height, and the span between anchors is no more than 3–4m, any fall will be minor and should be of no consequence. Danger exists if the traverse line is rigged too low or if cavers climb above it. A fall factor 2 (200 per cent) drop on to a cowstail in the middle of a traverse line generates loadings high enough to break the rope. Beware, also, of traverse lines that are rigged loosely, or with too great a distance between anchors. Any caver who falls on to a line rigged like this could be left dangling in an irrecoverable position.

At the pitch-head two bolt anchors are used to share the loading, forming a Y hang. Ideally these anchors are above head height, allowing the lifeline system to be rigged well above the pitch. Special care must be taken when connecting the lifeline system to the Y hang to ensure that the failure of either anchor will not have any catastrophic consequence. One way to achieve this is to use the bowline on the bight knot and then to connect the lifeline system to a karabiner clipped through (not between!) the two loops.

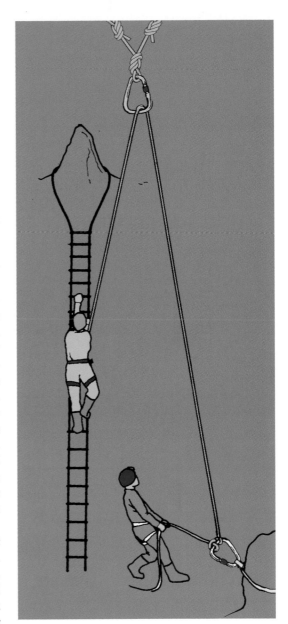

A double lifeline system protects the last caver down and the first caver up.

The ladder can be connected to any bolt in or out of the system that provides a good hang position and ease of access. For a technical rig with an ideally placed head-height anchor directly over the pitch and every caver using a harness, more sophisticated lifeline systems can be used, which have considerable advantages. If the Petzl Stop is used, lowering becomes an option, which is quicker and easier than climbing down the ladder. On the return climb assisting and hauling will be much quicker and more efficient than with an Italian hitch.

Abseiling

Climbing down ladders is strenuous and time-consuming, especially on pitches of 15m or more. Abseiling, a simpler and quicker alternative, is often preferred. There are methods that use the body as a friction brake (the 'classic abseil'), but these are extremely uncomfortable and are usually reserved for emergency use only. Most abseiling is by use of a mechanical friction device known as a descender.

Abseiling Accidents

These are becoming alarmingly common, and their causes include:

Abseiling off the end of the rope. This is easily prevented by simply tying a stopper knot in the end of the rope. This will prevent abseiling from the end but leaves the problem – now what?

Losing control. It may be fun abseiling fast, but it will certainly damage the rope and there is a danger of losing control. One metre per second is a sensible maximum descent rate.

Rigging the descender for minimum friction. Friction can be reduced on some descenders by an alternative threading pattern. This is useful on the upper half of long pitches, but only if extra friction can be applied mid-pitch as required, otherwise loss of control results.

Hung-up. This can be caused by a too short or tangled rope, or by hair, beard or other items being pulled into the descender and jamming. Prolonged suspension in a harness is potentially lethal, especially under a waterfall.

Accidents with the Stop Descender

While the stop descender has become standard equipment for British cavers, it is regarded more cautiously by many of our European neighbours who prefer the simplicity of the simple bobbin. According to British cave rescue statistics there is, on average, one serious injury every year caused by an abseiling accident using a stop descender. If we are going to continue promoting the stop as our descender of choice, it is essential that we reduce the number of accidents and to achieve that we need to understand their causes.

Problems

- One of the problems with the stop is that the mechanism is counter-intuitive. The instinctive reaction of the user descending too rapidly is to clutch the handle tightly rather than releasing it – the so-called 'clutch and plummet' accident.
- Another problem is its inconsistent behaviour with different ropes, some giving a smooth, easily controlled descent, while others are fast and jerky. This jerkiness can be so disconcerting that it causes the user to panic, resulting in a clutch and plummet.

Solutions

- Always use a braking karabiner to give more control.
- Always use two hands to control the stop descender. The right hand should grip

and control the slack rope entering the descender. The right hand is always first on and last off. The left hand should squeeze the handle only as much as is required to descend. With some ropes you will need to push the handle right in and control will be mainly with the right hand, but with other ropes just the lightest touch on the handle will allow a very smooth and controlled descent. (Some cavers take the view that the handle should only be used for on/off control, but it is my firm belief that this makes accidents more likely. It is also suggested that using the descender this way can damage the rope, but this has not been my experience.)

- Consider using a simple bobbin descender rather than a stop.

Precautions

- Use a selection of rope types and diameters during training.
- Use a lifeline, a safety knot, or lowered loop to protect the inexperienced stop user.
- Do not assume that other cavers are competent in the use of the stop until it is absolutely demonstrated.

Abseil Training

All of these potentially fatal errors can be avoided by adequate training, practice and preparation. It is a technique that should be practised on the surface while protected by a lifeline. It is a wise precaution, when a lifeline is used, to rig the abseil rope using a releasable system, such as a locked-off Italian hitch. In the event of a 'hung-up' problem, the abseil rope can then be released and the person lowered off by the lifeline. Abseiling must only ever be attempted using an approved harness – never from a belt.

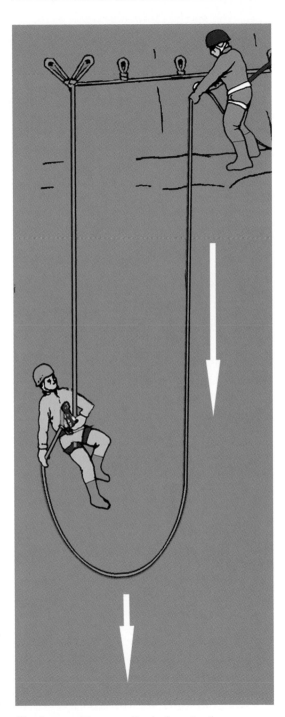

The lowered loop method of protecting an abseil descent.

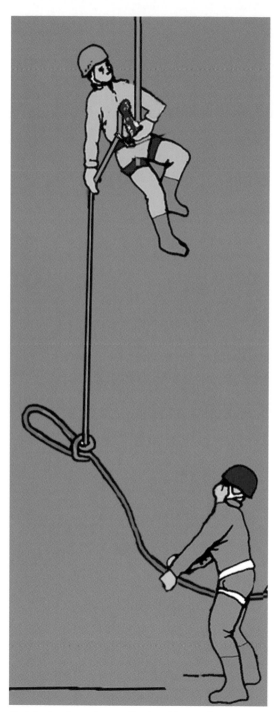

Safety knot system – the knot is easily released as the abseiler reaches it.

In view of these potential hazards, on-the-spot training for beginners is not advisable unless adequate safety precautions are taken. It is a safer option to anchor an inverted descender over a pitch and use this to lower down inexperienced cavers. Obviously the rope must be long enough, and the operator should have sufficient expertise to deal with any possible problems. This technique is often used by instructors as a quick and easy alternative to a ladder descent.

Abseiling Options

Italian Hitch

This is an extremely simple and effective method and gives a very smooth control-lable descent. It can cause – especially if several cavers descend by this technique – severe rope kinking and, especially if the rope is muddy, abrasion to the sheath.

Figure 8 Descender

Commonly used by climbers, this descender functions best on a double rope. On single ropes, particularly thin ones, descent can be very difficult to control unless additional friction is applied. It is useful, but not indispensable, for some pull-through trips. Not suitable for SRT.

Figure 8 descender and rope threading.

Rack Descender

The rack gives an extremely smooth descent and is well suited to very deep pitches with single drops of more than 100m. Friction is varied by the number of brake bars applied, and by adjusting the spacing between them. It is very important to apply more friction as the weight of rope below decreases, or a loss of control may result. The descender can be used with single or double rope.

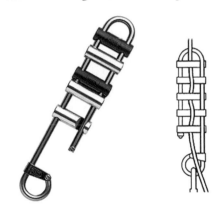

Rack descender and rope threading.

Bobbin

A single rope descender that gives a smooth controllable descent (if used as intended with a braking karabiner). Locking-off is quite complex and requires about a metre of rope, but otherwise it is a simple and effective device.

Stop Descender

This is the most popular caving descender. Since it has the benefit of an auto-lock mechanism, it should only be necessary to let go of the handle for the mechanism to engage and stop the descent. However, the auto-lock system does not always work, especially if it is well worn or on new, dry or thin ropes, so some caution is required. There have been several 'clutch and plummet' accidents using this descender, which emphasizes the importance of training and surface practice. It is also useful for lifelining and hauling.

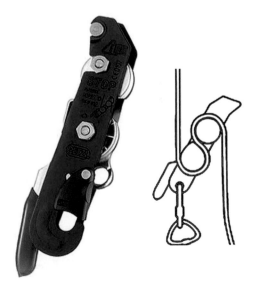

Stop descender and rope threading.

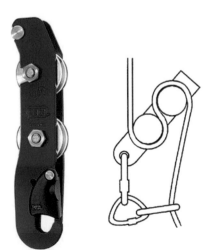

Bobbin and rope threading.

Self Lifelining

Double lifeline systems are not popular with cavers. They require a considerable amount of kit, are often difficult to arrange and frequently tend to jam. An alternative is the self lifeline system, which involves one person abseiling down a single fixed rope and then using a jammer for protection on the upward climb. The jammer can be fixed directly to the harness or attached to a cowstail, which enables it to be used as a mobile handhold. A climber falling due to a slip or tiredness will be held by the jammer until the ladder can be regained.

This can be a useful and safe technique in chimneys or free-climbable shafts where the caver's weight is easily transferred to the cave wall. But on a free-hanging or smooth-walled pitch, if the ladder fails, the climber could be left hanging helplessly in a harness, or, much more seriously, from a belt. The caver must, therefore, be able to effect a changeover from jammer to descender, a process most easily done using a second jammer and footloop. Effectively, the equipment required to safely self line a pitch of this type is a complete personal rig suitable for single rope techniques which makes any use of the ladder unnecessary.

CHAPTER 7

Single Rope Techniques (SRT)

In 1931 Dr Karl Prusik devised a simple knot that enabled climbers to escape from crevasses. The Prusik knot is still used and the process of climbing a single rope by the use of knots, or mechanical jammers, is known as prusiking. The idea of using single ropes for cave exploration has always been very appealing and first attempts date back to the 1930s in France (for an account see Pierre Chevalier's *Subterranean Climbers*). At that time the system was slow and painful, and the natural fibre ropes prone to rotting and sudden failure. It was not to be until the 1960s, in America, that SRT was to come of age.

Texan cavers, with numerous deep 'pits' to explore, devised the rope-walking system using Gibbs ascenders and non-twist nylon rope, which enabled them to climb hundreds of metres with relative ease. During the same period, British cavers were exploring Provetina, a 400m deep shaft in Greece, using first ladder and then winching techniques, in a major expedition requiring much manpower that ultimately succeeded in getting just two cavers to the bottom on separate occasions. In 1973 two Texan cavers bottomed the cave using SRT in a routine trip of less than seven hours. The message was clear – SRT had arrived and caving would never be the same again!

During this early period rope failure caused by abrasion resulted in several fatal accidents and the use of protectors and pads (as practised by the Americans) was urgently adopted. Meanwhile a new approach to SRT was evolving in the deep alpine caves of Europe. This alternative technique used multiple 12mm bolt anchors, which could be installed by hand in about twenty minutes. Bolts allowed a very precise line of descent to be taken away from water or other hazards

Jammers (self-locking rope clamps) are the basis of SRT.

Single Rope Techniques (SRT) – the easier way to get up and down. (Brendan Marris)

Learning SRT

It is normal practice to begin by learning how to follow a pre-rigged rope. The rigging of this rope is a separate skill to be learnt only after mastering the basics of getting up and down. Practice must start on the surface, preferably under the guidance of an experienced user or instructor. Various manuals and videos are available and can be used for reference, or even (cautiously) for self-training. It is a good idea to have some instruction before buying your own SRT kit so that you understand the functions of the various component parts.

Kitting Up

The basic SRT rig requires the following components:

1 × caving sit harness
1 × semicircle 10mm maillon rapide
1 × body jammer (Petzl Croll or similar)
1 × basic or handled jammer
1 × descender (rack, bobbin or stop)
1 × pair cowstails (2.5m of 10mm dynamic rope)
1 × footloop and safety connection (3.5m of static 8 or 9mm rope)*
1 × 3m chest strap or chest harness
2 × snap link karabiners
1 × screw gate karabiner (alloy or steel, oval or D-shape)
1 × oval screw gate karabiner (alloy or steel)
1 × steel screw gate karabiner (oval or D-shape)

* An alternative system using a separate dynamic safety connection is preferred by some cavers.

You may like to add to this list the Petzl Pantin foot jammer. This device makes progress up the rope considerably easier, but

and soon aid traverses and the techniques of rebelay and deviation became standard. This technical approach to SRT rigging was based upon the use of a very simple prusiking system: the sit-stand or frog method.

The 1970s in Britain were a period of experimentation as people devised different ways to rig, abseil and prusik. It became apparent that the very fast and elegant American rope-walking system did not combine with the complexities of European multiple bolt rigging and for a time, in Britain, the two systems coexisted uneasily together. Ultimately the European system was better suited to the multiple, generally short, wet pitches of British caves and that is the system in use today.

It is important to understand the historical background, and to appreciate the less obvious benefits of European SRT. Experimentation is all very well, but not if it just repeats the experience, and mistakes, of a previous generation, or reinvents the wheel!

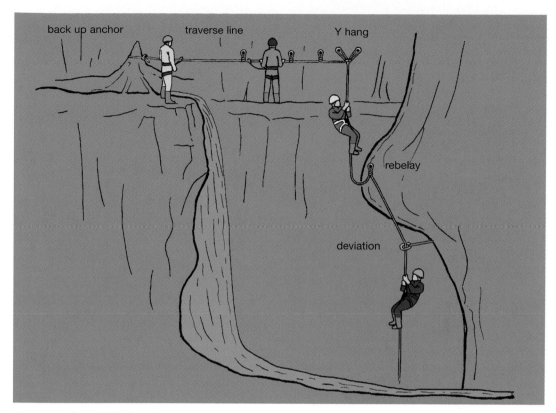

Features of an SRT pitch.

it is advisable to master basic progression and changeovers before you attempt to use it.

The following additional items are advisable:

1 × spare basic jammer
1 × oval karabiner 1 × fixed cheek pulley
1 × prusik loop (0.5 metre of 6mm cord)
1 × knife with retractable blade

Cowstails and footloop/ safety connections can be bought ready knotted, but since they must be adjusted to suit the user, there is often no advantage in this. Pre-tied knots also stretch as the rope is loaded and require much adjustment to return them to their ideal length. It is important to adjust the rig

to suit the user, otherwise prusiking is liable to be inefficient and basic changeovers excessively difficult.

There are some options in how you make up your cowstails. If you use figure 8 knots you will be able to easily remove the karabiners, but without metal clips or rubber keepers the karabiners have an annoying habit of not being correctly orientated when you need to get them clipped in. To make up cowstails of this type you need 2.5m of dynamic rope (I would recommend 10mm diameter). To get them the right length, it is best to start by tying a figure 8 knot in each end of the rope. The knots should have small loops and short tail (a stopper knot is not required if the knots are well tightened by body weight before use). A third figure 8

A typical SRT kit.

The precise arrangement of equipment is important.

Pantin foot jammer.

knot is tied next to one of the end knots, as close as you can get it (this may look wrong but it will stretch to the right length). Connect the end knots alternately to a head-height anchor point and bounce your weight vigorously on the third loop. Keep adjusting by reducing the length of all three loops and gradually tightening the arrangement until the knots are very secure. This should give cowstails of the right length, one about 30cm and the other about 75cm.

You can make your cowstails using knots of half a double fisherman's type, otherwise known as a barrel knot, and these will lock tight onto the karabiners, making them a permanent fixture with the correct alignment for easy clipping. The third knot in the system is usually a figure 8 in this arrangement, just over 2m usually being sufficient.

The footloop/safety connection is tied using an overhand knot at the harness connection end and a bowline loop (big enough for both feet) at the other end. A second overhand knot is tied, positioned to give the length as illustrated, and the footloop length is then adjusted until ideal. The knots should be loaded, stretched and readjusted until they are tight, with the lengths exactly right for the user.

Prusiking

Connect both jammers to the rope and sit down on to the harness. The chest strap will now be quite loose and needs to be tightened up so that the body jammer is held firmly against the stomach. The rope should be placed on the outside of one thigh while both feet are put into the footloop. The rope can then be placed between the feet and gripped. The hands are held in the position illustrated, ready to pull the body towards the footloop jammer. The feet should be brought right under the body so that the propulsion comes from beneath, and now it is only necessary to raise the body in a standing action.

Held firmly between the toes, the rope should slide easily through the body jammer. The next step is to sit back into the harness, open the toes to release the rope, relax the legs and raise the footloop jammer. Then the process is repeated. If there is sufficient weight of rope, or weight attached to the rope, for it to pull itself through the jammer, prusiking is easier and faster.

You can arrange a long prusik practice by suspending a pulley or descender 3 or 4m above the ground, and threading one end of a long rope through this. As one trainee prusiks up to the pulley, the rope is gradually released through an anchored descender.

Footloop adjusted to correct length.

Choose a chest harness that suits you – one that adjusts easily and makes your prusiking more efficient.

Prusiking: the sit-stand action propels the caver up the rope.

A stopper knot should be tied for safety. A proficient caver should prusik 50m in under seven minutes.

Reverse prusiking is a very useful technique that should be practised. The release of the jammers is done by pushing a finger down on to the cam and not by touching the release mechanism. Note that the jammer cannot release until it has lifted slightly and the teeth have disengaged from the rope. Reverse prusiking the body jammer is easiest if there is a distance of 20cm (equal to outstretched fingers) between it and the foot-loop jammer.

Abseiling

Beginners in SRT should consider using a simple bobbin descender in preference to an auto-locking model like the 'stop'. First use of the descender should be either protected by lifeline, or a short distance (maximum 2m) above the ground with a soft landing. If you are using a stop the golden rule is always to use both hands, the right hand feeding and controlling the slack rope, the left gently squeezing the handle as required. The right hand should always be first on and last off. There is only one correct way to thread the stop and it is easy to make a mistake. Incorrectly threaded, it still functions as a descender but has no auto-lock mechanism. In view of this, and the fact that some new, dry and thin ropes do not activate the auto-lock, particularly on worn descenders, it is important to use a safety lock-off when threading the descender. Ropes that do not auto-lock can be difficult to control and additional friction may be required. The spare steel karabiner should be used, placed next to the descender, on the user's right. With this extra 'braking' karabiner, control is much easier but locking-off becomes more complex.

Basic Changeovers

Being able to change direction on the rope, from abseil to prusik, and prusik to abseil, is a potentially life-saving skill. It is not uncommon for the first caver to descend and find that the route is too wet or the rope is too short. If water is the problem, and the caver's carbide lamp is extinguished, it may be necessary to changeover in the dark. The minimum skill level advised before a trainee ventures underground is that they are able to do basic changeovers. A competent SRT caver should be able to complete these with their light out or eyes closed.

Prusik to Abseil

Thread the descender onto the slack rope below the body jammer. Ensure there is a minimum length of slack between the two, and then lock off the descender. Lower the footloop jammer to 20cm above the body jammer. Stand up in the footloop, release the body jammer and lower your weight down onto the descender. Lower the top jammer again until it is just above the

Prusik to abseil changeover – threading the descender.

descender, but do not remove it yet as it is still your safety back-up. Unlock the descender and check it is correctly threaded. Remove the top jammer and descend.

Abseil to Prusik

Lock off the descender. Attach the top jammer and push it well up the rope, open the body jammer, stand up and clip it to the rope above the descender. One prusik up, while pulling slack rope through the body jammer, makes the descender easier to unthread.

If you find this method too strenuous there is an alternative. Clip the short cowstail to the jammer and abseil down until your weight transfers to this. It is then possible to remove the descender, and then clip the rope into the body jammer.

Knot Passing

Knot passes should be avoided underground by all but the most confident cavers, and only then attempted on dry pitches. But this is a good manoeuvre to practise on the surface and helps to develop general confidence and expertise. When ropes are joined mid-pitch, a clip-in loop is always left for cowstails connection; this can be simulated by simply tying a figure 8 knot (this needs to be at least 2m above the ground and at least 2m below the top).

Passing the knot upwards is not technically difficult. First, clip a cowstail into the loop for safety. Then remove the top jammer and replace it above the knot. Push the jammer well up the rope, stand up in the footloop, remove the body jammer and replace it above the knot. It then only remains to remove the cowstail and the operation is complete. Beware of pushing the jammers right up against a knot as they can completely jam in this position. This is a hazard to be aware of in many other situations, such as rebelays and pitch-heads.

To pass the knot downwards, abseil until the knot stops you. Change from abseil to prusik, prusik up once to give some slack, and then remove the descender. Reverse prusik down to the knot and you have two options: either do a normal prusik to abseil changeover with the knot between the body jammer and descender, or put a cowstail into the loop, transfer the body jammer to below the knot, and then complete the changeover.

The downward knot pass should indicate if footloop and safety connection are the correct lengths. It should also help to illustrate why these lengths are so critical.

The Rebelay

This is a point where the rope re-anchors mid-pitch, sometimes to a secondary Y hang, otherwise to a single bolt anchor. It can usually be practised on a convenient tree or rock face installed with suitable anchors. The rebelay loop should hang to a metre below the anchor point. Passing the rebelay upwards is similar to passing the knot: a cowstail is clipped into the anchor first for safety, but this time the body jammer should be transferred first. The advantage of this does not become obvious unless the belay above is significantly offset, or if there is good deal of stretch in the rope above.

It is very easy to get into a tangle when transferring the top jammer across. To avoid this, remove the feet from the footloop, disengage the jammer from the rope, and then hold the jammer at arm's length, with the footloop hanging free before connecting it to the rope above the rebelay. Do one prusik up, remove the cowstail, and then shout 'Rope free' if anyone is waiting below.

To pass the rebelay downwards, abseil down until level with the anchor and then lock off the descender. Clip the short cow-

A double bolt rebelay on a blank wall.
(Brendan Marris)

stail into the anchor, and the long cowstail into the rebelay loop, between the descender and the anchor. Unlock the descender and abseil until the weight transfers to the rebelay. Unthread the descender, transfer it to the next section of rope and lock off. Now it is necessary to lift the body weight off the short cowstail so it can be unclipped. Often this can be done by using available footholds, or by putting a knee or foot into the rebelay loop. An alternative is to use the footloop, clipped by a karabiner and not the jammer, into the figure 8 knot at the rebelay. Having achieved this, check the threading of the descender before removing the final cowstail. Remember to shout 'Rope free' if anyone else is waiting to descend.

Passing a rebelay downwards using both cowstails.

Passing a rebelay upwards. The body jammer has been transferred and the footloop is held clear before clipping onto the upward rope.

The Deviation

Deviations are often used on wet pitches to avoid water, and consist of a single cord and karabiner that deflects the rope into a new position. These are sometimes improvised by the rigger, using the most fragile of belays, in an attempt to stay relatively dry. Since they are so liable to failure it is very important not to confuse the technique for passing them with the method used for a rebelay.

The procedure, whether abseiling or

Approaching a single bolt rebelay in Nettle Pot. (Rob Eavis)

prusiking, is to stop just above or below the deviation, remove it from the rope and then clip it back into a new position either above or below you. A cowstail can be used to keep the deviation within reach but not for safety. If you do swing away from the deviation, and the rope is anchored or rebelayed below, it is possible to regain it by simply pulling towards it using the rope.

Traverses and Pitch-heads

These are difficult to simulate on the surface and usually have to be practised underground. The first experience should be as 'user-friendly' as possible and not a blank-wall aid traverse. Both cowstails should be used routinely and each one only briefly removed as anchor points are passed. The smaller the footholds the more aid is required, and the smaller the distance between anchors. The principle remains the same, but progress becomes more difficult.

At the hang point two anchors are usually rigged in equal tension to create a Y hang. Move as close to the Y hang as possible and then thread and lock off the descender. Clip the long cowstail into the lowest possible point, which is generally the Y hang, then take out the short cowstail and lower the body weight onto the descender. This should leave you hanging on the descender protected by one slack cowstail. Unlock and check the threading before removing the last cowstail and abseiling down.

The system is more variable on the return. Prusik up to the knot, stopping just before the jammers reach it. Clip the long cowstail in at the highest point you can reach (clip in the short cowstail, too, if possible). Stand up in the footloop, release the body jammer and transfer the weight onto the cowstails. If the short cowstail is not connected, pull your weight up using the traverse line until you can clip it in. Remove the remaining jammer and the changeover is complete.

Rigging for SRT

The objective for the SRT rigger is to achieve an abrasion-free and dry hang that is also user-friendly. Most caving in Britain is in well-used and documented systems where bolt and resin anchors proliferate. Generally it is a case of rigging along an accepted route using pre-existing anchors. The position of anchors, rebelays and the necessary rope lengths are often published in the form of topo guides, which are invaluable for inexperienced riggers.

Allow 2–3m of rope for each Y hang or a rebelay, 1m for every traverse line anchor,

then add a bit for good measure. Before going underground, the ropes must be end-knotted and fed into tackle bags in reverse order of use. The best position for the rope bag during rigging is to be clipped to the belt or hung from the harness maillon by a short cord.

Rigging often begins with a reliable natural anchor well back from the pitch-head. If rigging begins from bolt or resin anchors, ensure that two are used before traversing out over the pitch. While rigging a protection traverse line, the caver can be safeguarded in two possible ways: either by using cowstails clipped into loops tied into the rope, or by using a locked-off descender. If the traverse line is rigged too slack, a butterfly knot can be used for adjustment and tensioning.

Feeding rope into a tackle bag ready for use. The rope must be end-knotted first!

Jammers can be used for protection, but only if there is no possible risk of a shock load. No single length of shock-loadable traverse line should exceed 5m.

Some old designs of 8mm bolt hangers connect directly to the rope without karabiners or maillons and are well suited for use on traverse lines, but require all knots to be adjusted correctly before the bolt is tightened. Most riggers will find twist hangers a much easier option. Any further natural anchors can be rigged using either a sling, karabiner and butterfly knot, or by direct rope tie using a double bowline (this uses a good deal of rope). Resin anchors can also be used by a direct tie using a double bowline, which is extremely useful if maillons and karabiners are running short.

If the traverse changes from protection to aid, the anchors will be closer together. It is then possible to rig while clipped by the long cowstail into the previous anchor. Whenever rigging or de-rigging try to clip directly into loops tied into the rope. Be aware at all times of the consequences of any anchor failing, and minimize the risk by having at least two connection points.

Two anchors close together, often fixed into opposite walls, usually indicate the hang-point has been reached. A Y hang must be arranged between the anchors allowing an even distribution of the load, and a precise positioning of the rope. The ideal knot is the bowline on the bight because this adjusts easily between three points. It is an easy enough knot to tie, but it can be very frustrating to adjust correctly and really needs to be practised on the surface. Butterfly knots can be used, and these provide a simpler option. Angles of 90 degrees or less should be arranged on the arms of the Y hang as previously described.

OPPOSITE: Rigging clear of the water in Giant's Hole. (Rob Eavis)

The knot should be adjusted to hang the rope as clear as possible from the cave walls. When the rope is loaded all the traverse line anchors should, ideally, come into slight tension. This ensures there is maximum back-up to the system, with minimal possible shock loading. The rigger should descend cautiously, watching the rope above for any contact with the rock. Often, slight abrasion can be eliminated by a further adjustment at the Y hang. On some short, sloping pitches, keeping the legs outstretched against the wall will stop the rope from rubbing.

If an abrasion point is reached, a rebelay or deviation must be rigged. Usually, the position of available anchors determines which is to be used, but sometimes there may be a choice. If you are less than 5m down the pitch, a rebelay requires two anchors and takes the form of a second Y hang. If two anchors are not installed a deviation may be preferable. A deviation used to prevent abrasion should be rigged, using a sufficiently long cord, to clear the rope from the rock by only about 20–30cm. This minimizes the loading and ensures easy passing. Attaching the karabiner with an easily adjustable knot, such as a bowline or clove hitch, makes this easier.

If choosing between the deviation and rebelay there are several considerations. The rebelay uses more rope and is more difficult to rig and pass, especially for less experienced SRT users. It slows down the descent but should speed the party up on the return prusik. The deviation uses less rope, can often be improvised from natural belays and should, if the angle of deviation is not too acute, be easier to pass. But if the deviation is rigged too acutely it can be impossibly difficult for less experienced cavers to negotiate.

Rigging a Rebelay

A rebelay may use one or two anchors, but the rigging system always begins in the same way. Stop and lock off level with the anchor(s). If a resin anchor is in place, clip a karabiner or maillon into it. If rigging from a spit, use a plate hanger with a chain of two maillons or karabiners. Clip the short cowstail into the resin anchor or upper maillon and abseil down until the body weight transfers to this. Feed rope through the descender until it goes slack and then lock off again.

Now tie a figure 8 knot about half a metre (up to one metre if a double anchor is to be used) from the descender and clip this into the remaining maillon. Continue, passing the rebelay in the usual way. If two anchors are to be used, tie a butterfly knot below the figure 8, connect this to the second anchor, and adjust to a 90-degree angle. This is much easier than attempting to use a mid-pitch bowline on the bight. Using this system should ensure that just enough slack rope is left to easily thread and lock off descenders at the pitch-head above.

Connecting Ropes

Pitches often follow on directly from each other and sometimes the end of one rope is used partially or completely to rig the following pitch. It is very important that any rope-ends long enough to have a descender mistakenly threaded on to must be end knotted for safety. Surplus rope should be left, suspended in a tackle bag or coiled and hanging from an anchor. Where ropes join on a traverse they should be directly connected together for safety.

My only recommended knot for a mid-pitch rope join is a double fisherman's knot. Be very wary of using a re-threaded figure 8 as this knot is prone to failure if the ropes are of different diameter. The double fisherman's knot should be tied, leaving a sufficiently long end for a figure 8 safety loop. This makes a very secure join but can be very difficult to untie after loading. Another

Different methods of joining SRT ropes together mid-pitch. In both cases the join includes a loop for clipping into for safety. The system on the right uses a reef knot between the double fisherman's to make untying easier.

option is to tie a reef knot, then the double fisherman's, and then to join the ends together with a second double fisherman's. This is very easy to untie, provides a good-sized safety loop and is extremely secure.

Pendulum Swings

There are sometimes ledges or windows partway down a pitch that give access to a better or drier hang, or even to a whole new series of passages. A pendulum swing may be necessary to reach the right spot and this can be initiated by pulling or pushing against the cave wall or by using an improvised grappling hook (a tackle bag on the end of the rope works well). It is important to keep a close eye on the rope above during these manoeuvres as serious abrasion is a hazard if the rope rubs.

Guidelines

An effective way of avoiding water on a wet pitch is to use a tensioned guideline. This is

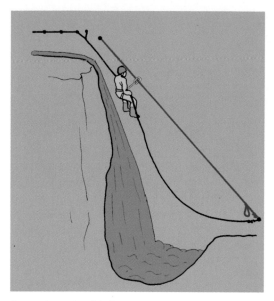

A tensioned guideline can be used to avoid water.

rigged at an angle and tensioned using a Z-rig system. Cavers who are abseiling or prusiking the pitch clip a cowstail (preferably using a pulley) into the guideline, which then pulls them clear of the water. Some caves, such as the viciously wet Diccan Pot, have guidelines that are left permanently in place.

Rope Protectors

These are simple wraparound pads secured with Velcro and held in position with a Prusik loop. In the early days of SRT it was the standard method for preventing rope abrasion. The rope protector still has a role to play during prospecting descents, particularly at or near the pitch-head, but it is awkward to use mid-pitch, and often slips from the intended position. The use of multiple bolt anchors throughout Europe has made the rope protector virtually obsolete for most cavers.

Rope-Walking

Sit-stand SRT is much less efficient on big pitches than the rope-walking systems developed and still used in America. Conventional rope-walking does not combine easily with European rigging techniques involving rebelays and deviations. It is possible to arrange a hybrid rig that includes the benefits of both systems. This is sometimes referred to as the fantastic-elastic system. The body jammer is fixed in a slightly higher position (usually using a long maillon to link it to the central maillon), which raises the centre of gravity and keeps the body upright. The footloop jammer operates at about knee height, connected to the shoulder by an elastic cord. The footloop must either be easily adjustable, or have two footloops of different lengths. A third jammer is also used (a Petzl Pantin is ideal), fixed directly to one foot. A safety karabiner connects the central maillon to the rope, and will stop against the footloop jammer in the event of body jammer failure or disconnection.

Used in this way, rapid and easy progress is possible. If rebelays or deviations need to be passed, the footloop jammer is transferred to its normal position (a longer footloop is then required, hence the need for quick adjustment) and conventional techniques are used. This is a system for very experienced users who are already entirely proficient in the standard technique. It only really becomes essential when ascents of very big pitches (more than 200m) are planned.

A simpler version of rope-walking, the frog-walk, can be achieved with a standard SRT rig and a Pantin foot jammer. You need to master a different sequence of movement to achieve the rope-walking technique. Try it – it does work when you get the knack!

Cord Technique

Cord technique is a system that reduces the amount of rope needed to rig deep caves. A thin cord is used to lower the rope down and then to hoist it back into position, enabling it to be used for multiple pitches. Obviously such a system is fraught with potential problems and hazards and is not recommended for general use.

The 'travelling' rope is prepared carefully before use by trimming, melting and sewing back the sheath at one end until it tapers smoothly to a 3mm tail, 30cm long. The other end of the rope is tied into a standard ball-shaped stopper knot. The pitch-head is rigged in a conventional way using a separate rope, and a 7mm maillon rapide is anchored at the hang point.

For the descent, the 'travelling' rope is fixed, running through the maillon, and held in place by the stopper knot. Attached to the stopper knot is a length of 3mm cord twice the length of the pitch. When the team has descended the rope, the other end of the cord is tied to the tapered end and the rope is then retrieved by pulling down on the cord connected to the stopper knot. This leaves the cord in position, ready to pull the rope back up when required.

It is essential that the cords do not twist or tangle. The ends should be linked together and the two sides anchored well apart. Deviations can be used on the pitch but rebelays are not practical with this system. If deviations have been used, only one side of the cord will pass through them, so it is vital to know which side is which.

On the return, the tapered end of the rope is reconnected to the cord and it is hoisted up through the maillon until the stopper knot jams into position. It is advisable to connect the other end of the cord back to the stopper knot during this operation, so that if the system jams, retrieval is possible.

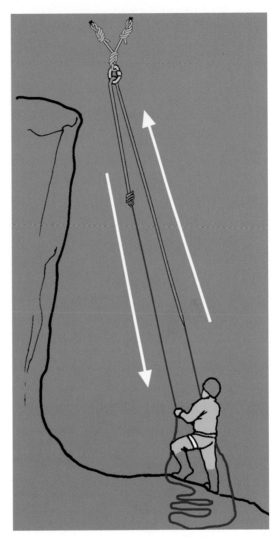

Retrieving the SRT rope by using a thin cord for hoisting and lowering.

The pioneering Australian caver Alan Warild is an exponent of this system and his manual *Vertical* has much useful advice. Cord technique is usually only relevant to the ordinary caver in two situations. It is sometimes used to provide access to high-level passages (but normally without the refinement of the tapered rope-end) and it has applications in the next topic, the pull-through trip.

Pull-Through

The most satisfying caving trips are, without doubt, through-trips, especially those that follow the water right through a system, from sink to resurgence. Any pitches en route can be descended by abseil, and the rope then retrieved by pull-through. These trips are very popular and often appeal to cavers with limited technical skills, who are not always sufficiently aware of the potential hazards.

Pull-through trips are entirely committing and once the first rope has been pulled down there can be no retreat. Also, because the rigging cannot be technical, many descents are under torrential waterfalls. Rope jams are common and, if no spare rope is carried and no group member is equipped to prusik, marooning between pitches is possible.

Any caver who leads this type of trip should be aware of its potential hazards. The releasable abseil rope system (using a locked-off Italian hitch) is recommended, but this is of limited use without sufficient spare rope in reserve (to allow lowering to the pitch-bottom), unless the abseiler is connected to a separate lifeline.

Pull-through around natural belays is often unsuccessful as excessive friction or tapering cracks can cause rope jams. It is usually necessary to use a sling or short rope around the belay, and pull through this. Rope-to-rope contact is very abrasive but can be avoided if a 7mm maillon is sacrificed. It is very wise to check that the rope will pull through before the last caver descends.

The safest way to arrange a pull-through trip is for every team member to be SRT proficient and equipped, for spare ropes to be carried, and for single rope pull-down systems to be used. Even experienced cavers have been known to connect to the wrong rope using this technique with serious consequences, and it is best to keep the pull-down side coiled or bagged until the last person descends.

Old slings and cords abound at some pitch-heads, many of which are unsafe, but the most popular multi-entrance cave systems are often rigged for pull-through, with double bolt or resin anchors. When rigging for pull-through, it is essential that the failure of any anchor will not be crucial. A short rope tied with three figure 8 knots to form a Y hang (one to each anchor and one for the

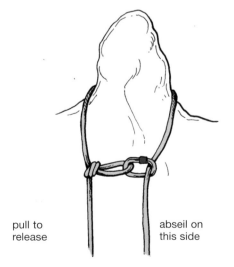

pull to release   abseil on this side

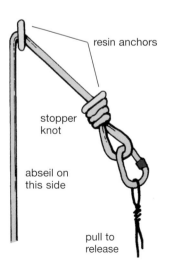

resin anchors

stopper knot

abseil on this side

pull to release

Pull-through methods.

abseil rope) is the easiest way to achieve this.

Resin anchors are often placed close together, one vertical and the other, below and offset, at 45 degrees. This enables the rope to be threaded through both anchors for safety in a way that reduces friction during pull-through. A stopper knot is sufficient to secure the rope when using resin anchors, and it is possible to reduce rope lengths by joining at the stopper knot, or even to use cord technique for the pull-through.

Pitch-head safety is easily compromised on this type of trip. It is still quite practical to rig short traverse lines to the pitch-heads, or even to arrange that these can, themselves, be pulled through by threading through various anchors. It is true that you 'should never trust your life to a single bolt', but it is much wiser to clip a cowstail into a handy resin anchor while preparing to descend, than to be clipped to nothing at all.

Problems

There is one crucial difference between SRT and ladder and lifeline techniques: the caver connected to a lifeline can easily be assisted by those above, but the caver connected to a single rope cannot rely on the help of others. There are, actually, many ways to assist someone stuck mid-rope, but often only by the use of complex, and hazardous, techniques. Prolonged suspension in a harness has a tourniquet effect on the femeral arteries that can result in unconsciousness and death within twenty minutes. A caver suspended under a waterfall is equally at risk from rapid hypothermia. This is why SRT demands a high level of personal proficiency before use underground, and also why every SRT caver needs to be prepared for possible problems and emergencies.

This type of training cannot usually be provided by the caving club alone and is most effectively taught by a specialist instructor. It is a subject that is never entirely learnt because there are an almost infinite number of subtly different situations that can develop. It is a foolish and selfish caver who considers it unnecessary to attend a training session where his friends are taking the time and trouble to learn how to help him in an emergency.

Exhaustion

This is one of the most common problems and is linked not just to general fitness, but also to power-to-weight ratio. Some cavers find basic SRT changeovers extremely difficult because they rely so much on arm strength, and do not have a free hand to release the body jammer easily. The difficulty is increased if they have a heavy upper body and they find it strenuous to keep in the upright prusiking position. A solution to this problem can be contrived by using a 2:1 prusiking system in which the footloop cord connects to the body jammer via a pulley clipped to the upper jammer. The French manufacturer Petzl produced a very neat version of this called Le Pompe, but sadly it never really caught on. The benefit of these systems is that they take much of the strain off the arms and make mid-rope manoeuvres much easier.

The 2:1 system can be improvised with a standard SRT rig, provided it is arranged in the manner previously described. The long cowstail should first be clipped to the jammer as a substitute safety connection, and then the footloop unclipped from the oval karabiner. This is clipped back again with the safety connection running freely through the karabiner and the system is now converted to 2:1 advantage. Prusiking becomes slow, but much easier. A pulley can be used to make the system slightly more efficient.

If a tired caver is making very slow

progress up a pitch while you are waiting at the top, it may be possible to provide assistance. If you have a rope long enough to reach him, lower down one end for them to connect to a cowstail. Clip a karabiner or pulley into the Y hang, run the rope over this and down through your body jammer. It is possible to use one capstan of a stop or bobbin descender as an improvised pulley. You can then use your body weight suspended on one side of the rope to counterbalance the caver and make their progress much easier. Keep your own cowstails connected for safety while doing this.

Improvised Equipment

What do you do if you drop or lose the use of a vital piece of gear such as a jammer? If you routinely carry a spare jammer there should be no great problem, otherwise you will have to improvise. Here are some useful techniques:

Italian hitch. This is a simple substitute for a descender that most cavers are familiar with. Practise using it on changeovers and simulated rebelays.

Prusik loop. This works adequately as a replacement for the upper jammer, but does not function easily as an emergency body jammer.

Garda knot. This simple system (also known as an Alpine shunt) uses two identical karabiners (avoid HMS or large diameter D-shaped) to provide a simple self-locking knot. It works well but should be used with caution, as it cannot be released while under load. This can be an effective substitute for a body jammer but changeovers to abseiling should be practised.

Descender used as jammer. It is possible to use a stop descender as an emergency jammer. The stop works well when threaded around one capstan. Progress is slow and strenuous, but should get you there in the

Garda knot.

end. The stop also has the benefit of being rigged for instant descent if necessary (the rope should be held over the top capstan and a braking karabiner used).

Hazards and Emergencies

Imagine the situation – a rock is dislodged and falls on to a caver prusiking up the pitch below. He hangs unconscious on the rope and will die in minutes unless released from his harness. Picture a flooding cave or a caver hanging, hopelessly tangled, in the full force of a waterfall, vainly trying to changeover. You could be watching this drama from above or below, and you could be their only chance of survival.

Rescue Training

The priority should always be to get the casualty off the rope by the safest and simplest method available. There are several possible options, depending on the available kit and the rigging of the pitch. The position of rebelays and deviations may be crucial in determining which system will work.

These techniques will only be effective if they are familiar through regular practice. Attempts to apply these methods with only partial knowledge could result in the would-be rescuer becoming entangled with the victim and, in the worst case, this might cause a double fatality. The following descriptions are included as an aid to training. The DVD *Speleo-Vertical* illustrates these techniques in considerable detail and is strongly recommended as another training aid.

Mid-rope Cut

This is the best option unless the pitch is excessively wet. A spare rope, long enough to reach the pitch bottom, and a knife are required. Anchor the rope at the Y hang (it can be tied directly through the loops) and abseil down to the casualty. Stop with about a metre (harness to harness) above them and lock off using a braking karabiner. Connect your long cowstail to the casualty's harness, and his to yours. Very carefully cut the casualty's rope just above his head. Warning: tensioned ropes cut very easily – be very careful while the knife blade is exposed. Before abseiling down with the casualty you must end-knot the cut rope-end to prevent another accident.

Casualty on Auto-lock Descender

If the casualty is hanging from an auto-lock descender, the knife will not be necessary and another option becomes possible. Abseil down, as before, on the spare rope, but this time connect to each other using short cowstails. If the spare rope is long enough to reach the pitch bottom, lower the casualty's weight onto your harness by releasing their descender. If the spare rope is not long enough, abseil down, transfer your weight to the casualty, and abseil using their descender (with a braking karabiner).

Should it be necessary, the same principle can be applied to change ropes again and used to avoid rebelays. Alternatively, rebelays can be untied, and then abseiled straight past as abrasion is not likely to be a hazard during one abseil descent. Worry about the problems of prusiking up later – just get the casualty down quickly and safely.

Pitch-head Cut

If the spare rope is not long enough to reach the casualty, but does exceed the distance between casualty and pitch bottom, this is the best option. Clip an inverted descender into the Y hang. Tie a figure 8 knot in the end of the spare rope and connect a jammer to this. Connect this (upside down) onto the casualty's rope, about a metre below the Y hang. Thread the spare rope through the descender and lock it off. Now push the jammer down the rope to take up the tension (and to check it's the right way round). Very carefully cut the casualty's rope between the jammer and the Y hang, unlock the descender and lower them down.

The success of this system is dependent on the rigging of the pitch. A rebelay above the casualty will make it unworkable, while a rebelay below will not prove a problem unless it is higher than halfway between the floor and the casualty. A little thought about the subject will clarify this. Ask yourself another question: what effect will deviations have, above or below the casualty? Remember that the rope may be bottom anchored, and that an end knot in the rope cannot pass through a deviation.

Pitch-head Lift

If you have no spare rope it may be possible (again this depends on the rigging of the pitch) to lift the casualty. If someone is in difficulties under a waterfall it may only be necessary to raise them a few metres to a drier position. Clip a chain of three or four karabiners, or a short sling, into the Y hang. Attach a jammer to this and clip it, upside down around the casualty's rope. Direct lift or counterbalance lift are both possible. For direct lift release your chest strap, allowing the body jammer to turn upside down, and connect it, inverted, on to the rope. Lift using the thigh muscles and gather slack through the jammer. Alternatively, transfer the footloop jammer from the usual overhand knot to the footloop itself. Run the safety connection through one of the karabiners in the chain or, better, through a pulley clipped to the Y hang. Now use your body weight to counterbalance the casualty. This works best if you, or somebody else, attaches another jammer or prusik to the rope to pull upwards on.

When sufficient slack has been gathered it is possible to clip a pulley (or a pulley jammer) into the Y hang and run the rope through this. Counterbalance can then transfer to the slack rope emerging from the pulley, and continue more efficiently. With more slack gathered it becomes possible to use a Z-rig to lift the casualty. Remember that your priority is to take the casualty's weight from the harness. The greatest problem, if the casualty is unconscious, may not be the lifting but getting them off at the pitch-head.

The Double Cut

If you have no spare rope but you do have a knife this is the best option. You will have to use the casualty's rope to reach them by prusiking up or reverse prusiking down. If you are approaching from below you need to get above them by transferring your jammers over theirs. Once in position this is the sequence. Disconnect the casualty's upper jammer from the rope and from any foot-loop or safety connection. Clip this jammer to the rope above you. Cut the rope just beneath the casualty. Tie the cut end into a figure 8 knot and clip this to the jammer you placed on the rope. Thread your descender to the rope and transfer your weight to this (locked-off). Connect a short cowstail to the casualty. Cut the casualty's rope just above their body jammer. Unlock and abseil down.

Mid-rope Rescue

This is the final option when nothing else is possible (and all because you don't have a knife). It is difficult and potentially very dangerous for the rescuer, who risks becoming hopelessly entangled with the casualty if things go wrong. It must be practised regularly to stand a reasonable chance of success and it is a wise precaution to have some way of easily and safely aborting during practice. Reverse prusik down to the casualty and connect long cowstails together for safety. Remove the casualty's top jammer and prusik down again until you can clip his short cowstail to the underside of your central maillon. If possible, convert your prusik system to 2:1 advantage. Now prusik up, lifting the casualty until it is possible to reach down and release his body jammer.

Thread your descender, using a braking karabiner, and lock off. Prusik again using the 2:1 system until you can release your body jammer. Allow the overhand knot to pass through the oval karabiner if necessary (this is when you realize why an overhand knot is recommended). It now only remains to descend with the casualty.

If the casualty is above you it will be necessary to pass him mid-rope. Prusik up to him and connect cowstails for safety. Transfer your top jammer over his body jammer and then remove their top jammer from the rope. Transfer your body jammer over his and the operation is complete.

CHAPTER 8

Hazards

Is caving a dangerous activity? Ask a non-caver this question and the answer will probably be a resounding yes. It is almost as if a propaganda machine has been set to work against us. Cave rescues have always made better news stories than cave discoveries and are a recurring theme on docu-dramas of the '999' type, complete with horrifying reconstructions. If that was not bad enough, dramatists regularly seize upon caving to create cliff-hanging plot devices. A flood or a rockfall is no longer sufficient – it must be both. As the public watch their heroes squeezing between perilously balanced boulders, while the surging floodwater rises higher by the second, it is not surprising that their view of caving is, to say the least, tainted.

In reality you (and your children) are much less likely to sustain an injury caving than you are playing football, mountain biking or riding a horse. I cannot provide you with statistics to prove this assertion, but I can say that after thirty-seven years of caving I have never personally known anyone who has died through misadventure underground. Amongst the hundreds of cavers that I have known a very, very few have suffered minor injuries, in every case through a minor slip or fall.

Parties led by qualified instructors or highly experienced club cavers very seldom get into trouble, which illustrates the simple fact that most cave rescues are caused by inexperience and/or human error. Rescue reports and statistics illustrate well the types of incidents or problems that do occur underground. I would recommend taking a long and careful look at the incident reports on the British Cave Rescue Council website (www.caverescue.org.uk).

Throughout the rest of this chapter I will list the most common causes of accidents and cave rescue call-outs, along with advice and guidance on how these might be avoided.

Overdue

Overdue parties can cause a lot of anxiety to friends and family members and can be very disruptive for the members of the rescue team. It is quite a regular occurrence for rescue organizations to 'scramble' a team, sometimes late at night, and for these unfortunates to tramp across a sodden moor just in time to see a group of unscathed, if rather embarrassed cavers, emerging from a hole.

Sometimes these debacles result from inadequate or confused call-out information; for example, by not clarifying the difference between 'estimated time out' and 'call-out time'. More often, though, the cause is a failure to allow sufficient time for the planned trip. Exiting from a cave usually involves an upward climb for increasingly tired cavers and it should be no surprise that this takes longer than descending. A party that has used 40 per cent of its allotted time, but is still continuing downwards, is almost inevitably going to be overdue.

The timings of through-trips and round-trips can be quite problematic, especially when, having passed the halfway point, the route becomes complex or more difficult. To complete this type of trip within an allotted time you really need to have navigational aids, realistic expectations of your party's abilities, and to seek advice on the probable duration of the trip from someone familiar with it. Also, when it comes to timing, it is not helpful to realize partway through the trip that nobody in the party has a watch.

Lost on Surface

There can be few cavers who have never experienced this! Cave entrances close to the road or on footpaths are easy enough to find in daylight and clear visibility but emerging later into darkness and mist on an isolated moor can be a real problem. Come out in a blizzard and the surface becomes more hostile than the cave. Anticipate the event and carry a compass routinely. If you are caught out, get a fix on any wind direction and use that as an indicator.

Lost Underground

Even relatively short and easy caves can present route-finding difficulties and these are massively increased in a really extensive multi-levelled system. Route planning should begin well in advance of a trip into labyrinthine networks like Easegill Caverns or Ogof Ffynnon Ddu. Guidebook descriptions are helpful but much less so than a survey, which is the crucial aid for navigation. If you relate the text description to the survey well in advance both should make more sense. Having a copying and laminating capability is very useful for creating cave-proof surveys.

Inexperienced cavers are sometimes so focused on finding the way onwards that they fail to note the way back. It should become second nature to look back at junctions and memorize any features. This is very important if you drop into a passage or chamber from a climb in the roof or emerge from a small, concealed hole in the wall.

Another problem here is the tendency to write off possibilities. When a group is looking for a way on or way back, the correct route may be probed for a short distance until a squeeze is encountered or another junction is unknowingly passed. The caver concerned, having missed the right way, or assumed, wrongly, that the squeeze is too tight, declares the route to be wrong. This decision goes unchallenged and no further search is made in that direction. So always double-check, and if necessary check again.

Falling

Historically, the greatest single cause of death and injury in British caving has been falling, unlifelined, from ladder pitches. Better techniques, equipment and improved training have now combined to reduce this type of incident. New equipment needs new skills from the user, however, and misuse can be highly dangerous. There have been several accidents caused by incorrect use of sticht plates, shunts and, in particular, auto-lock descenders. The problem has been rooted to a large extent in the learning and training culture of many individual cavers and their clubs, which has often lagged behind the new realities of the equipment revolution.

Other falls are less avoidable. The terrain underground is seldom level and often slippery. Short climbs and slopes abound and any minor slip or fall can result in injury. Novice cavers are always most at risk. Do not hesitate to rig handlines or lifelines if you feel they are necessary, regardless of whatever

Waterfalls and wet pitches quickly become deadly barriers during floods. (Rob Eavis)

the accepted practice may be for a particular climb. Physical support, a helping hand or a foot held on to a foothold is often all that is required. 'Spotting' is a term that instructors use to describe the practice of assisting and supporting other cavers on short climbs. It means standing in a secure position with hands outstretched ready to support a slipping person – it may mean you both take a bit of a tumble if the climber slips but you are a softer landing than limestone and two bruised people are better than one injured one!

Flooding

Flooding is the single biggest cause of cave rescue. In some regions flood rescues are almost routine, and on any wet Saturday a party is likely to be in difficulties somewhere. The chances are that the incident will be in a popular cave with a well-recognized flood danger. The fact is that many cavers simply do not appreciate the danger or know how to assess the risk. There is also a dangerous notion, commonly held, that wet conditions are more 'sporting'.

Almost any cave with a stream will flood to some extent. Guidebooks are often inconsistent in their references to flood liability: some even happily use terms like 'refreshing' and 'exhilarating' to describe high water conditions in specific caves. It is always safest to assume that any cave with an active streamway is floodable unless there is specific evidence to the contrary.

Not all floods begin with rain. Some cave systems are affected by irregular discharges

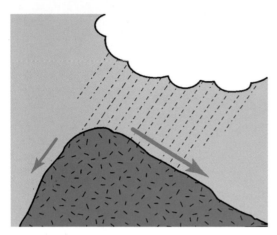

The rain shadow effect can concentrate flooding on hillsides that face the prevailing weather.

from reservoirs, such as Goyden Pot and other systems in Nidderdale, North Yorkshire. Another danger here is that, under some conditions, a strong wind can blow dangerous amounts of water over the dam. Sea caves are well known for their tidal dangers, but only one major British system, Otter Hole in the Wye Valley, is subject to this hazard. One of the original exploration teams had a traumatic near-miss in this cave before its tidal nature was fully understood.

The most likely reason for dry-weather flooding is snow melt. What is in effect several days' worth of rain can lie in the form of snow drifts, the extent of which are not always obvious in the valleys below. The danger is at its highest when the snowfall is followed by mild drizzly weather. This can gradually saturate the snow until it forms a slushy mass that can release large quantities of water quite suddenly.

When rain does fall the true intensity is not always apparent from a valley floor. Rain shadow effects often mean that the hillside facing into the prevailing weather (in Britain usually south to west facing) receives much more rainfall than slopes aligned in other directions. The steeper and better developed

the catchment the more efficiently it will conduct water to the cave. Meregill Hole, in Yorkshire, illustrates just how efficient a catchment can be. It is situated in the centre of a huge natural, west-facing funnel. The upper catchment, is nearly vertical and a system of well-developed stream beds converge on the cave entrance, each capable of delivering a very sudden and substantial volume of water.

The flood danger in some moorland areas has been increased by the dubious practice of 'gripping'. In an attempt to improve the grazing, systems of drainage ditches have been excavated. A herringbone pattern of shallow gullies now converge upon many stream sinks and these can have a profound effect on the flooding of caves.

The same amount of rain falling can have quite different effects depending on existing conditions. During the spring and summer a lot of moisture is absorbed by plants and vanishes quickly into the bio-system. Topsoil can absorb and store a huge volume of water, depending upon its level of saturation and the intensity of the rain. Heavy rain falling upon very dry ground will have a high run-off, but the danger is worse when the ground is entirely waterlogged after prolonged wet weather. In these conditions, when no more moisture can be absorbed, run-off is at its maximum.

Streams in flood carry vast amounts of debris and sediment. If a stream entering a cave is gin-clear you can reasonably assume that it is not exceptionally high. If it is even slightly cloudy, discoloured or, in peaty moorland areas, has the appearance of black tea or beer, stream levels are already above normal.

High water levels underground mean danger. Progress is more strenuous and communication difficult, while conditions are colder and much more hostile. Hazards are increased and the result of an accident is

Foam is an indicator of the level of flood waters. (Rob Eavis)

likely to be much more serious in the hostile conditions. Any ensuing rescue operation will be substantially more difficult and dangerous for everyone concerned.

A cave in full flood is a potentially lethal environment. A cubic metre of water weighs one tonne, and a fast-flowing stream needs to be only knee-deep to sweep a person away. Deaths have occurred in these conditions. Falling water is especially dangerous and presents a serious barrier, which is good reason to rig ropes and ladders as far from waterfalls as possible.

As flood water sweeps through a cave it will free-flow through any unobstructed sections, and back-up or pond before constrictions and blockages. The free-flowing water will sweep away existing sediments and scour the passage, leaving clean rock surfaces behind it. In the ponded areas, sediment and debris settle, leaving deposits of mud, debris and foam on the floor, walls and even the roof.

With experience, these indicators become quite obvious. Passages such as the Main Drain in Lancaster Hole clearly illustrate these principles at work as lengths of scoured streamway alternate with areas of sediment deposit. Some areas show both features in the form of scoured floors and clean walls up to a distinct level, above which there is thick mud with fragments of debris. These areas are subject to both effects as flood levels rise to a peak and then fall.

In caves where the passages are constricted, these indicators are inadequate as relatively small amounts of water can be dangerous. These systems often have small catchments that can respond quickly to a sudden downpour. In a narrow passage the body can act as an obstruction, causing water to build up dangerously in front of the caver in a situation where just a few centimetres may be enough to cause drowning.

As a flood reaches the lower sections of a cave system it can cause a general rise in the water table or phreas, which can result in the submerging of both active and fossil passages. In Lancaster Hole many passages are flooded as water backs up from the master cave sump, including some in other more distant areas that have phreatic links.

Flash Floods and Flood Pulses

Rainfall is generally predictable in Britain unless thunderstorms occur. Thunderstorms concentrate extreme quantities of rainfall on to small areas, often with dramatic results. The intensity of rainfall can generate a flash flood or flood pulse that will surge down watercourses, both above and below ground, as a wave or even wall of water. The only warning of this event may be minor flood pulses entering through inlets or a sudden increase in draught.

Comparative Rainfall

Continuous drizzle	around 1mm per hour
Light rain	around 2mm per hour
Continuous rain	around 4mm per hour
Heavy driving rain	around 6–8mm per hour
Thunderstorm rain	25–100mm per hour

How to React to a Flood

The first reaction will be an urgent desire not to be in the cave, but do not react to this by

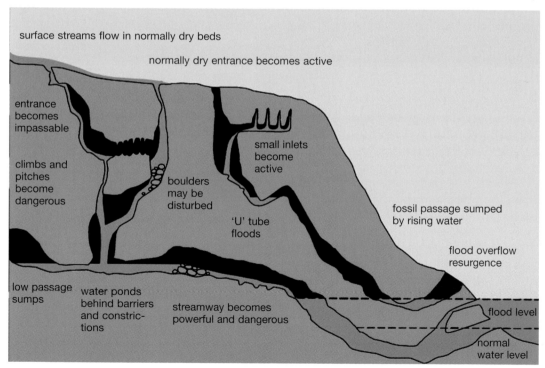

The effects of flooding.

making an immediate dash for the entrance! Stop and think carefully about the cave. Which areas are likely to flood, and which are safe? High-level passages, especially those with unmuddied formations, are usually not flood-prone. If there is no immediate danger, or when you have found a safe area, monitor the water levels to see if they are still rising, stable or falling. Now is the time to appreciate that balaclava. Stay warm by huddling and using emergency bags or shelters. Be patient: most floods subside within a few hours. You will not die of boredom or starvation. The embarrassment of cave rescue and media attention may be painful but it will not kill you.

Weather Forecasts

All this drama can be avoided by obtaining, and acting upon, an accurate and recent weather forecast. Recent means within six hours of the intended start time, and this does not include newspaper reports collated the day before. The Radio 4 morning forecast at 7.55 a.m. is sufficiently detailed, or there are weather call telephone services that are very localized. The most convenient and accurate source of information is online at sites like www.metoffice.co.uk or www.bbc.co.uk/weather. Using these resources you can get a very accurate hour-by-hour prediction of expected rainfall.

I have cancelled many trips into floodable caves because of adverse forecasts, often being, in retrospect, over-cautious. On the eve of a Pen y Ghent Pot trip the forecast was bad and things looked doubtful. Next morning one of the team (who was very keen to do this cave) reported that the forecast was much improved and so we entered the Pot. Four hours later, as we struggled back against raging flood water, I determined always to check the forecast personally in future.

Continental Caves

These observations mostly apply to the British climate and cave systems. In mountainous areas there is often very little vegetation or topsoil and run-off is consequently always very high. The summer climate is such that a daily cycle often occurs with massive convective cloud build-up in the late afternoon followed by dramatic thunderstorms. These storms, which are highly dangerous on the surface, cause an almost daily flood pulse in the cave systems. In this type of region the only reliable way to avoid flooding is to cave during the night and morning.

In areas such as Scandinavia where snow melt is an important factor, warm dry weather may cause more severe flooding than rain. In tropical regions, caving is often restricted to the dry season, but in some areas this is only a relative term and lethal flooding remains a constant danger.

Boulders

It is in the nature of limestone to fracture along joints and bedding planes and to separate into square or rectangular blocks. Caves are subject to the process of breakdown by which sections of roof and walls peel away and collapse to the floor. Most large passages and chambers are strewn with

Boulder slopes are often unstable and must be negotiated carefully. (Steve Sharp)

a chaos of tumbled rocks that makes for tedious and troublesome going. Sometimes the route is entirely filled with a choke or 'ruckle' of boulders and the only way on is cautiously to follow a devious three-dimensional route between them.

Nature has left many of these boulders in a precarious balance – 'hanging-death' waiting for the careless or unwary caver to put a foot or hand wrong. Where the bedding is steeply inclined, such as the Mendip Hills, boulder slopes are very common. There is a real danger on boulder slopes of disturbing a loose rock that could endanger either you or cavers below you. It really is important to think about every move, pick the safest route and avoid climbing directly above or below others. In the boulder choke the main route will often be polished by passing bodies and this can be used as indication of the way through.

Chockstones, which are rocks wedged between the passage walls, can be dislodged by people climbing over, or crawling under, both of which have caused accidents. One of the worst accidents in British caving happened in Easegill Caverns in 1988 when three cavers, who were off-route in a boulder ruckle, were killed by a massive collapse.

Even small stones become deadly missiles when falling just a few metres. Climbs and pitches often abound with loose rock. Extreme care is required not to dislodge material at the top and not to be a target at the bottom – always keep well clear in a sheltered position. 'Gardening', the making safe of loose material, is often required above a pitch or climb before a safe descent is possible. It is important to warn of falling rock by shouting 'Below!', and not to react to this by looking up.

OPPOSITE: Treading lightly on precariously balanced boulders. (Chris Howes)

Bad Air

The natural ventilation of caves usually ensures that the air underground is fresh. Carbon dioxide is an occasional hazard most commonly associated with cave digs, as it is heavier than air and collects in low unventilated areas. It is usually noticeable that the air seems stale and that normal progress is more tiring than it should be. Carbon dioxide can be detected by using a match, candle flame or miner's flame safety light. If the flame is reluctant to burn, levels are high, and a prompt return to fresher air is advised. Note that carbide flames cannot be relied upon to conduct this test.

The situation is aggravated in some caves below farms, where oxidizing sewage not only increases the carbon dioxide level but reduces available oxygen. This type of cave usually shows other indications of pollution such as worms, flies and a strong 'farmyard smell'. It is usually best to enter this type of system in the winter, or after a wet spell has flushed most of the pollution away.

There are some caves that have high carbon dioxide levels without any apparent reason. In the Mendip Hills this has become an increasing problem, especially in the summer months, that sometimes affects major cave systems. Bad air is a known problem in some tropical caves, probably because of the large amount of rotting vegetation carried into caves as flood debris.

Other Hazards

Leptospirosis

Leptospires are bacteria, often found in freshwater streams, that can cause serious illness. The organisms live in the kidneys of various host animals and are then excreted into watercourses, where they can survive for several months. The organism survives most readily in clean water, rather than polluted or deoxygenated conditions.

The most severe form is spread by rats and causes the illness known as Weil's Disease. This develops after about ten days and usually begins with severe flu-like symptoms, possibly with diarrhoea and muscle paralysis. Jaundice then develops and the condition is fatal for about 10 per cent of patients. Only prompt diagnosis allows for effective antibiotic treatment. Most doctors have little experience of this condition, and it is important to insist that a blood test is taken urgently for analysis.

A second form of leptospirosis infection is carried by cattle. This variety is less serious and causes milder symptoms, but can persist for up to six months. The patient is liable to feel run down and generally off-colour, but will eventually recover completely. The infection is also a problem for cavers overseas: 'Mulu Fever', an illness commonly affecting cavers in Sarawak, has been identified as a form of leptospirosis.

Prevention. Caves taking farmyard, industrial or domestic drainage are most likely to be affected. The infection enters the body through the mucous membranes of the mouth and nose or through cuts or grazes. Gloves should be worn routinely if a risk is suspected. Any existing wounds should be covered in a waterproof dressing, and any fresh cuts thoroughly cleaned and disinfected. Drinking cave water is never recommended in caves subject to contamination. Eating and smoking should be avoided during and after the trip, until the hands have been thoroughly washed.

A Code for Safety

20 Golden Rules

1 Choose a cave suitable for the whole party.
2 Eat before caving and carry spare food.
3 A minimum of four cavers is sensible.
4 Carry a map and compass in open country.
5 Ensure that every caver is well equipped.
6 Leave a written message with someone responsible.
7 Assume that any active cave can flood in wet weather.
8 Do not try to fight flood water.
9 Carry spare lights and emergency equipment.
10 Memorize junctions and the correct route back.
11 Practise ropework techniques on the surface first.
12 Select anchor points carefully.
13 Move cautiously on boulder slopes and pitch-heads.
14 Do not swim without flotation aids.
15 Never attempt to free-dive an unknown sump.
16 Enter descending squeezes feet-first.
17 Do not consume alcohol before caving.
18 Save time and energy for the exit.
19 If in doubt, quit while you still can!
20 Teach novices safe practice, and set a good example.

Radon

Radon is a naturally occurring, colourless, odourless radioactive gas that can deposit carcinogenic particles in the lungs. It is widely recognized as a potential health hazard, particularly in regions where it accumulates in homes and buildings. Studies have revealed that some caves have seasonally high levels of radon, especially those in the Castleton area of Derbyshire.

Radiation doses are measured in millisieverts (mSv). Legislation permits workers an exposure of up to 15mSv per year, unless registered as radiation workers, in which case a limit of 50mSv per year is permitted. The highest known accumulations in British caves are in Giant's Hole during the summer months, and a regular visitor to the cave could well exceed a dosage of 50mSv during this period. Tests suggest that other regions in Britain have much lower levels, though isolated and possibly temporary 'hot spots' may occur.

The health implications of radon in caves are most significant for those who work underground, such as caving instructors and showcave guides. Many showcaves have now installed ventilation systems and reduced the time that staff spend underground.

CHAPTER 9

First Aid, Survival and Rescue

Safe practice and good judgement should minimize the hazards encountered underground, but can never remove them completely. We should have no illusions that danger will always exist to some extent, and that accidents are possible. In any cases of serious injury or hypothermia, the initial actions of those on the scene can make the difference between life and death or, perhaps, between full recovery and permanent disability. Being able to administer essential help often depends upon carrying a small number of useful items.

First Aid, Survival and Emergency Kits

Group Kits

In the 'formal' led group, the instructor or leader is responsible for providing an appropriate first aid and emergency kit. Assuming the group numbers between six and ten persons, this might include the following:

1 × pair disposable gloves
1 × roll insulating, canoe or gaffer tape
1 × triangular bandage
1 × large sterile unmedicated dressing
2 × medium sterile unmedicated dressings
6 × medi-wipes
1 × knife
2 × packets glucose tablets
1 × large, heavy-duty survival bag
1 × balaclava helmet Waterproof writing material, with rescue call-out procedure reminder
1 × waterproof torch Chocolate
1 × metallic mug Solid fuel tablets
4 × candles ('night-light' type are best)
1 × cigarette lighter

This kit is specifically intended for the cave environment and the probable emergencies that might arise. It does not include ordinary sticking plasters, since these are of little use in wet or muddy conditions. Adhesive tape is much more effective for closing cuts or securing dressings, and can also be useful for repairing electric lights or carbide hoses.

The kit is very much geared to survival, and to the prevention and treatment of hypothermia. One solid fuel tablet will heat a mug full of water to near boiling point, but only attempt this in a well-ventilated, or spacious, area. Glucose tablets dissolved into hot water will help to raise or maintain core temperature and blood sugar levels. Candles can function not only as an emergency light, but also as a heat source. An effective way to light a candle or solid fuel tablet is with a disposable cigarette lighter, which is stored sealed inside a plastic bag.

The kit is best carried in a completely sealed waterproof bag or, better, in a robust plastic drum. It's a good idea to check after every trip that the vessel is still effectively waterproof or you may find your emergency kit a sodden useless mess when the time comes to use it.

Personal Kits

The ordinary caving trip is often informal with no individual taking responsibility to provide a group first aid kit. It is advisable for the individual caver to be self-sufficient and to carry a few useful items on their person. This is an extension of the personal emergency kit with some items serving a dual role. It could include:

1 × roll insulating tape
1 × medium sterilized unmedicated
    dressing
1 × lightweight survival bag
1 × balaclava
2 × candles ('night-light' type are best)
1 × cigarette lighter
1 × knife
Chocolate or glucose tablets
1 × waterproof torch

Some of these items are easily carried on the person in undersuit and oversuit pockets (only soft items can be carried safely inside the helmet). A small plastic bottle can carry a surprising amount, and is easily hung from the belt or carried in a bag.

First Aid for Cavers

There is only one way to learn the basics of first aid adequately, and that is by a recog- nized training course. For the caver, special- ist mountain first aid is much more relevant than the standard domestic or workplace course. Some caving clubs arrange suitable courses, particularly those with links to cave rescue. Certificates are usually valid for three years, but most first aiders would recom- mend an annual refresher whenever possi- ble, especially if this includes resuscitation, which requires regular practice. Remember, your friends may be taking the trouble to learn to help you in an emergency; don't you owe the same to them?

Following an Accident

First assess any danger. If the casualty has been hit by falling boulders, be reasonably sure that the area is safe to enter. If they have slipped down a steep slope, it may be best to secure a rope before descending yourself. Do not rush in and become another casualty.

The first action if the casualty is uncon- scious is to check their airway. Noisy, rasping or gurgling breathing indicates an obstruc- tion. Open the mouth and check for vomit or broken teeth and clear if necessary. Hold the tongue clear of the back of the throat or, if neck injury is not suspected, tilt the head back using the chin.

Under normal circumstances the next pro- cedure would be to call an ambulance, but if you are underground this may present you with some hard choices. If there is only you and the casualty, do you leave an uncon- scious person alone while you summon help or allow their condition to deteriorate fur- ther while waiting several hours for a rescue triggered by your call-out? If there are three of you, should one of you solo out and if so who?

If the casualty shows no indications of breathing (or is not breathing normally), pinch the nose and cover their mouth with yours (or a CPR mask). Blow into their mouth and notice if their chest rises. If so, give two breaths, each lasting one second.

If the victim is still not breathing normally, coughing or moving, begin chest compressions. To be effective the chest must depress at least 38 to 50mm. Press down on the chest on a line running from nipple to nipple, in the centre of the chest. Pump at the rate of 100 per minute. Continue with two breaths and thirty compressions until help arrives or you are too exhausted to continue.

This is the current procedure and differs from that recommended until quite recently. Resuscitation techniques are constantly being refined and improved, which is why it is important to attend regular training and refreshers. Whether or not you are confident in the technique, it is essential to at least try. Remember that a patient in cardiac arrest (who has no pulse) is 'clinically' dead. No further harm can be done attempting to resuscitate a person who is already technically dead.

It is important to be realistic about the likely effectiveness of resuscitation underground. Cardiopulmonary resuscitation (CPR) is only a temporary measure to keep the heart and brain cells alive and a spontaneous return to life is extremely unlikely. Normally a defibrilator, used by paramedics or other suitably trained person, is necessary to restart the heart. The chances of successful resuscitation of a patient in cardiac arrest, in a caving environment, are therefore extremely unlikely. The safety of the rest of the group, the locality and likely rescue time are therefore extremely important considerations in starting or continuing any resuscitation attempt.

The Unconscious Casualty

While awaiting rescue the casualty must have an airway maintained. The simplest way to

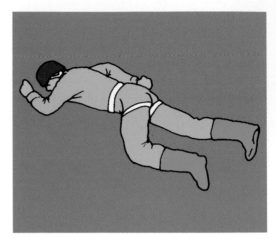

The recovery position.

ensure this is to place them in the recovery position, which keeps the tongue clear and allows fluids to drain from the mouth. However, if damage to the neck, spine, pelvis or body is possible, it is unwise to move the casualty unnecessarily. The size of the group is a crucial factor in these circumstances. If one of only a pair, the casualty may have to be left alone, and must be placed in the recovery position first. In a larger group the airway can be maintained with a chin lift and pulse and respirations monitored continually for deterioration. Maintaining the airway must take priority over all other considerations.

Head Injury

Head injuries are insidious because they may be more serious than they initially appear. Remove the patient from the cave promptly, encouraging them to move by themselves unless their condition prevents this. Talk to them and monitor their mental state for increasing confusion, amnesia, slurring or stumbling. Clear fluid leaking from the nose or ears, bruising under the eyes or behind the ears, or pupils of different sizes indicate a serious head injury requiring urgent medical attention.

Fractures

A fracture underground is likely to cause severe pain and immobilization requiring a cave rescue call-out. Generally the fractured limb should not be moved, and this is very important if the neck, spine or pelvis is affected. If the casualty is in a dangerous location, or at risk from hypothermia, it may be necessary to move them to a safer or drier position. Lifting should use all available manpower to provide maximum support and to maintain the existing body position. The head should be held in line with the body to prevent twisting of the neck. Pain can be reduced by gently supporting a fractured limb.

Asthma

This is a very common condition and an attack underground is potentially very serious. It should be standard practice for cave leaders to ensure that asthma sufferers take a Ventolin inhaler underground. There is little danger of overdose and the inhaler should be used until the symptoms ease. Used directly into the mouth, only 15 per cent of the Ventolin reaches the lungs. It is much more effective to use a spacer between the inhaler and the mouth. A plastic drinks bottle with a hole pierced in one end can be used if available, otherwise cupped hands are the best option.

Severe Chest Pains

A crushing pain in the chest may indicate angina or a heart attack. Rest the patient at once. If the symptoms ease, angina is most likely (the patient may carry medication if previously diagnosed), but if the condition continues a heart attack is probable. A heart attack is the only situation where the first aider is advised to offer a drug to the patient, in this case one aspirin, which should be chewed rather than swallowed for the most rapid absorption into the bloodstream. Sit the patient up in a comfortable position, and keep them warm and reassured while sending for help.

Bleeding

Reduce bleeding by direct pressure on the wound (gloves advised) and whenever possible by elevating an affected limb. Adhesive tape can be used to close wounds, but ensure that circulation to the limb is not seriously impaired.

Harness Suspension Trauma

This is a potentially lethal effect of being suspended in a harness. It is most critical if the victim is or becomes unconscious. Consider a 'hung-up' caver so hopelessly tangled that they are awaiting cave rescue. They will continuously shift their weight to ease their discomfort. These movements are actually essential. If the caver becomes exhausted and inert they will fall unconscious in as little as 10 minutes and unless removed urgently from the rope the condition will be fatal.

A casualty successfully removed from the rope must *not* be laid down but must be sat upright for at least 30 minutes. This is essential in order to avoid 'reflow syndrome'. Because the blood in the legs has been unable to flow during suspension it has effectively become toxic, and if allowed to flow back to the vital organs it can cause damage or death.

Hypothermia

Hypothermia is defined as a lowering of the body core temperature from the normal

37°C to 35°C (98.6°F to 95°F), or less. The body reacts to this loss of body heat by diverting available energy to the vital organs of the chest. Blood flow to the skin and limbs is first reduced, but as the condition worsens, and especially if the patient is exhausted, the functioning of the brain is affected.

This may cause behavioural changes such as sudden aggressive outbursts, or it may simply result in a lowering of the level of consciousness. Mental deterioration can be hard to recognize in someone who is already miserable and withdrawn. It is important to talk to the patient, to encourage a response, and to assess their condition constantly. Talking can serve another purpose by raising the patient's morale. The idea that some people can survive through sheer determination, while others lose the will to live and consequently die, should not be dismissed. In any emergency try to be cheerful and optimistic, no matter how worried you are about the situation.

If physical exertion continues, the muscles are effectively diverting the remaining energy reserves from the vital organs. Exhaustion and hypothermia are often linked together in a deadly spiral. It is essential that the hypothermic patient is rested, insulated and fed with glucose or sugar. The condition usually develops over a period of hours, during which the following symptoms may be observed, usually in this order:

Initial/Mild Hypothermia Symptoms
- Intense shivering
- Lack of coordination
- Poor judgement, stumbling, slurred speech
- Shivering decreases

Advanced/Severe Hypothermia Symptoms
- Shivering ceases
- Mental deterioration, incoherence, confusion

- Pupils dilate, pulse weak and irregular
- Unconsciousness

Early diagnosis and appropriate action are essential. A patient showing severe symptoms must be stabilized. The crucial warning sign is the stopping of shivering, or severe mental deterioration. This is an indicator that all physical exertion should cease, that the patient should be insulated, and rescue called.

Treatment of Mild Hypothermia

Give the patient additional insulation. Spare, relatively dry and pre-warmed clothing is available if group members are wearing long-sleeved thermal vests under fleece suits. This type of garment, worn routinely, can be considered a vital part of your emergency kit. If the layer of clothing directly next to the patient's skin is thin, and already warmed by body heat, it can be left in place. Wool and cotton garments should be thoroughly wrung out to remove water before being replaced. Heat loss through the head is very significant and can be reduced by using a balaclava or hat.

Give chocolate or glucose and reduce energy expenditure by lots of physical assistance. Assisted handlines are useful and can be arranged very quickly by using a human belay (the loadings are relatively low). Avoid waiting in cold or draughty areas. Try to raise the patient's morale. Monitor their condition carefully – are they starting to stumble, their speech becoming slurred, can he answer questions rationally? If they stop shivering be certain that this is due to re-warming, and not because their condition is worsening.

Treatment of Severe Hypothermia

Lay the patient horizontally and keep them in that position until help arrives. Do not remove their clothing and avoid any unnecessary movements. Insulate the patient from the ground using bags, ropes or, best, a human body. Give them a balaclava, wrap them in any spare clothing, and place them in a survival bag. Get spare people alongside to provide more warmth. Give glucose, if possible in a warm sugary drink. Organize a rescue call-out urgently. Keep talking to and reassuring the patient. If the patient loses consciousness ensure their airway remains open.

Survival

There are several possible reasons why a prolonged wait may be necessary underground. Parties can be trapped by fallen rocks or hopelessly lost, but flooding is the most likely cause. The trapped party should try to find a dry comfortable area that shows no indications of flooding and is as free as possible from draughts.

Every member of the party should be carrying their own survival bag and this can now be prepared for survival mode. The bag should be carefully unrolled to reveal the sealed end. One corner of this can be cut off, to leave a face-sized hole of about 15cm diameter. The oversuit can now be removed and folded to provide an insulated seat, while the survival bag is placed over the head and body, leaving the face exposed through the hole. The helmet chin-strap can be used to seal the bag around the face, or a balaclava can be worn over the bag.

Now you need a carbide lamp or candle to act as a heat source and fill the bag with warm air. Sitting on the oversuit with the knees up provides the best position. A carbide generator, held between the thighs, will help to warm the blood flowing through the femoral arteries and this can keep the feet warmer. As the air warms, the undersuit will

begin to dry and condensation will collect on the inside of the bag. By carefully taking off the bag and reversing it, it is possible gradually to dry underclothing. It is also possible, using the bag this way, to stand up, walk around and exercise cramped limbs.

To stay warm you really need to huddle up with other cavers, which is not always a very appealing prospect. If a large heavy-duty bag has been included among the group equipment a shelter can be improvised. The bag can be cut into a single sheet and suspended between dry stone walls. A bed of bags, ropes and oversuits should be arranged – remember that no amount of insulation on top is effective if you are losing heat through the

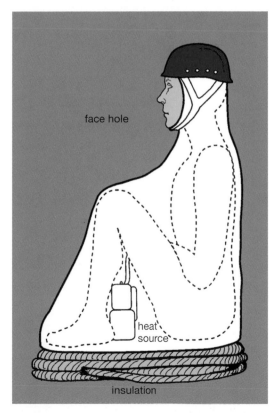

Position in a survival bag, insulated from the ground by rope and using a carbide lamp as a heat source.

Cave rescue is often an extremely difficult and arduous operation.

floor. A concentration of bodies and carbide lamps underneath will generate warmth and create condensation. Turn the sheet over now and again to alleviate this.

Ration your carbide, candles and any food you may have. As long as you can stay reasonably warm, hunger is of secondary importance. You will not starve even if trapped for a matter of days, though food may become an obsessive topic of discussion. If a small billy-can, or even just a tin mug, is carried, water can be warmed over a carbide flame or with a solid fuel tablet (beware of fumes). Warm water will maintain the body core temperature and help to prevent dehydration, which is a hazard in this situation.

Mountain Shelters

Walkers, climbers and some cavers now routinely carry mountain shelters, which are larger and more elaborate versions of the simple survival bag. The shelters are water- and windproof and can be spacious enough to accommodate several people. A typical four- to six-person shelter weighs less than 800 grams and packs down very compactly. This is a very effective way to combine body heat and to have the benefit of a psychologically 'cosy' environment.

Rescue

Rescue Call-out

In Britain cave rescue is the responsibility of the police, and voluntary rescue teams operate on their behalf. Cave rescue call-out is made by dialling 999 and asking for the police, and then asking the police for cave rescue. The police will take initial details and then contact the relevant cave rescue organization. It is essential that the caller stays at the phone until called back by the rescue warden or controller as they will need detailed information to ensure that the right equipment and personnel are dispatched.

Mobile phones are a mixed blessing. They do not always work well in caving regions and this can lead to time-wasting attempts to find a spot with better reception. It is not very helpful to borrow a mobile for an initial call if the owner then drives away from the scene before the rescue warden calls back. The most important point is that the mobile networks have central emergency control rooms serving large areas and may not immediately know which rescue team to alert. It is very important to ask for the relevant county police force.

Cave Rescue
- Phone 999 and ask for Police.
- Ask Police for Cave Rescue.
- Stay at the phone until contacted by the rescue warden.

Cave Rescue or Self Rescue?

A group of cavers are on their way out of a cave. They reach a ladder pitch that one member of the party, who is exhausted, cold

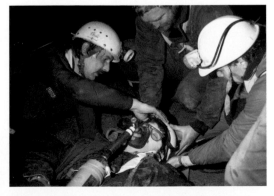

Regular training with specialist equipment is an essential element of cave rescue. (Chris Howes)

and demoralized, is unable to climb. The decision is made to call cave rescue. The rescue team will arrive as promptly as they can, but it can be a case of hours rather than minutes, especially if the emergency is deep underground and the cave remote from the nearest road. This leaves ample time for a case of mild hypothermia to degenerate into something much more severe.

Imagine an alternative scenario, where the group apply basic improvised rescue techniques, using slings, belts or a rope-end to improvise a harness and then haul using a simple 2:1 system. The team evacuate the exhausted caver from the cave, giving a push and a pull up every little climb with constant positive 'banter' and encouragement to help energize and lift the spirits.

Another option for the team is to send a caver or cavers to the surface to call for cave rescue while they begin a self-rescue evacuation.

You must not attempt self-rescue if:

- The casualty is incapable of self-mobility.
- You do not have the technical skills to hoist or reverse a hoist.

It was recently reported that an attempt to self-rescue an unconscious person ended as a fatality. This emphasizes that, while minor problems can be dealt with by self-rescue, more serious situations demand the specialist skills of a rescue team. When in doubt, call for rescue.

The Rescue Teams

Every caver would do well to attend at least one rescue practice. It takes approximately ten times as long to move a stretcher through a cave as it does for an able caver to travel the same distance. To manoeuvre a heavy stretcher through a constricted cave passage without aggravating existing serious injury is the rescue team's greatest and most

Annual statistics of rescue call-outs (British Cave Rescue Council)									
	1998	1999	2000	2001	2002	2003	2004	2005	2006	2007
Cave rescues	41	36	34	18	19	21	22	21	32	27
Persons assisted	70	47	60	27	41	41	38	17	65	38
Fatalities	1	0	2	2	2	0	1	2	3	3
Persons injured	11	6	12	6	9	9	9	4	3	6
Animal rescues	13	14	12	7	12	18	17	15	13	9

exacting challenge. First-hand experience of the difficulties of cave rescue gives the caver a realization that there are some locations, deep underground or beyond extreme squeezes, where extra care and safety precautions are essential.

Cave rescue relies entirely on cavers volunteering their services to help other cavers in trouble. There are more than a dozen separate teams covering caves and disused mines throughout Britain and Ireland, each with their own specialist knowledge of the sites in their locality. Some of the busier teams operate as a club in their own right, while others organize and coordinate training, practices and actual rescues through the local caving clubs.

Each team has a number of caving doctors, but they are unlikely to be first on the scene, so initial first aid and paramedic care is often a responsibility of the ordinary rescuer. It is therefore essential for the modern rescue team to provide regular high-quality training. There are so many situations that can arise underground that each team needs a comprehensive store of specialist equipment. The list includes ropes, pulleys, various stretchers, pumps, hydraulic lifting kit, Induction telephones, hot-air kit, and many other items, each of which needs training and practice to use effectively.

Each team publishes an annual report detailing rescue call-outs during the year, and all are represented on the British Cave Rescue

Distribution of cave rescues over a six-year period	
Yorkshire Dales	128
South Wales	61
Mendip	51
Derbyshire	36
Forest of Dean	11
Ireland	4
Devon	1
Scotland	1

Distribution of rescue incidents over same period	
Falling	60
Overdue	51
Flooding	36
Lost	30
Exhaustion/exposure	29
Physically stuck	18
Rockfall	12
Medical problem or illness	12
SRT related	9
Other, including diving	36

Specialist stretchers make progress through awkward passages easier. (Greg Jones)

Council (BCRC), which publishes overall statistics. The ordinary caver is well advised to study rescue reports as these often contain valuable lessons. You can find the British Cave Rescue Council website, including incident reports, at www.caverescue.org.uk.

The main causes of rescue are parties overdue, falling or flooding. The number of overdue incidents is greater than the figures indicate as not all teams record surface standbys in their annual figures. Certain caves are well known for taking much longer than cavers anticipate and when parties are overdue in these systems there is not an immediate panic to organize a search. It is important to realize this, as some cavers have chosen to remain with an accident victim because they expected the rescue team to arrive shortly after their call-out time.

Falls can usually be divided into two categories. There are those caused by slips on short drops that are normally climbed without equipment, and there are those associated with pitches. Almost every caver has a minor slip and tumble at some time and it is often just bad luck when injury results.

Falls from pitches are entirely avoidable. The biggest single cause of death and injury in British caving is falling unlifelined from ladder pitches of less than 10m. Another growing problem is the misuse of technical equipment. There have been accidents caused by the incorrect use of some lifeline systems, and through loss of control when using stop descenders for abseiling. All of these accidents can be avoided by education, training and practice.

Flooding, exhaustion and hypothermia are frequently related. Again, there is very little reason for this type of rescue to occur, other than poor judgement and inexperience. This is further indicated by the fact that most of these rescues happen in a small number of popular caves with a well-recognized flood danger.

In the public imagination physical jamming is perceived to be the routine hazard for cavers. This is largely because of the media obsession with the word 'stuck', which seems to be used when reporting any type of cave rescue. Most jammed cavers free themselves before the rescue team arrives, but sometimes there is a genuine problem and the team resorts to enlarging the passage and even smearing the victim with washing-up liquid. No British caver has died solely as a result of being physically jammed in a squeeze.

CHAPTER 10

Cave Leadership

Leader Qualifications

Certification of leaders has no part to play in ordinary recreational caving, which, in Britain, has a tradition of being unencumbered by regulation and bureaucracy. But, as with all adventurous activities, parents and employers often request some formal check of proficiency. The British Caving Association (BCA) operates two schemes to fulfil this requirement: the Cave Instructor's Certificate (CIC) and the Local Cave/Mine Leader Assessment scheme (LCMLA).

CIC is an appropriate qualification for those passing on technical skills to other cavers and instructing in a variety of caves at different technical levels. LCMLA is intended for leaders supervising groups within specific named caves, selected according to their personal experience.

The ordinary club caver has no requirement to seek these awards, but often finds the associated training courses of great benefit. The awards are based on a checklist of knowledge and techniques that are essential to every experienced caver, making this training element very relevant. The methods, responsibilities and ethics of leadership are also examined, and can give the ordinary caver much to reflect upon.

Role of the Leader

Most caving parties have one person who, by virtue of personal caving experience, or by familiarity with the intended route, takes on the role of leader. The safety of the group, the success of the trip and the enjoyment of those involved may depend entirely on the judgement and conduct of this one person. Many accidents and rescues result from just one bad decision by the leader.

Some groups have no leader and often consist of equally experienced cavers working informally together. Many clubs or groups operate successfully and safely in this way, but problems can arise when one weak or novice member joins the party. There can be much reluctance to be the one who says 'we should turn back', or 'you must use a lifeline'. Bear in mind an observation made by psychologists: people acting as a group are prepared to take more risks and hazardous options than they are as individuals.

Being a keen and capable caver does not automatically make an individual a good or safe leader. Judgement is the foundation of leadership and this is a skill that develops more gradually than learning to tie knots, or becoming the fastest caver in the club. A good leader must correctly assess the party's abilities, choose a realistic objective and be prepared to adapt to changing circumstances.

A leader should be sensitive to the morale, confidence and abilities of the party. Is a trip successful if it has reached the objective and returned safely, but one member is so cold, exhausted and dispirited that their previous enthusiasm for caving is permanently terminated? And what lessons are being taught underground? Are novices shown what an emergency kit should contain, are hazards and features explained to them, does the leader set an example by using a lifeline?

Many cavers progress quickly to personal proficiency, and begin leading club trips or taking their friends underground. Some may find that their considerable caving experience has not prepared them for the very different responsibilities of leading a group.

The following sections offer some advice on leading groups at various levels.

Insurance

It is a sad fact that we live, increasingly, in an age of litigation, when any minor mishap is seized upon greedily as an opportunity to profit. It is important for all cavers who lead others to be insured, and an inexpensive and comprehensive scheme for clubs has been arranged by the British Cave Research Association. Litigation is bad news for everyone except lawyers, but it does at least give the cave leader constructive pause for thought. These questions should be considered: can you justify all your actions, is your equipment at an acceptable standard, and have you taken all necessary safety precautions?

A novice group is thoroughly briefed before the underground adventure. (Andy Sparrow)

Novice Groups

Before taking any group underground a leader should ensure that novice cavers have clear instructions – preferably written – regarding suitable clothing and footwear. Those under eighteen years old should have written parental consent. There should be some explanation of what the trip will involve, and the novices should be asked if they have any medical conditions, such as asthma, that might affect them underground.

It is a function of the leader to provide moral support and gentle persuasion to an uncertain novice hesitating at the cave entrance or first low section, but nobody should be forced to enter a cave against their will. Caving does not appeal to everyone, and all prospective members of the group should have a realistic idea of what is involved and not be put under pressure to participate.

It is the role of the leader to reassure the group and to stress the relative safety of the activity, but this should not be overstated. Novices must be aware that they are personally responsible for their own safety and conduct themselves in a sensible way. Caving in the right location, under correct supervision and using suitable equipment is less hazardous than many more popular sports, but it is wrong to assume that it is entirely without danger or that all responsibility can be delegated to the leader.

The size of the group is an important factor. Large groups make slow progress and individuals are unlikely to have close personal attention. Slow-moving groups result in bored, restless and sometimes cold cavers, as well as causing overcrowding in popular caves. The optimum group size varies from cave to cave, but a maximum of eight novices with two leaders is a reasonable upper limit. Numbers need to be reduced if any ladder work is involved, as this can be very time consuming,

and often involves waiting in draughty areas.

The cave chosen should be well known by the leader. The ideal site has a complex of stable passages allowing a diverse route through a variety of obstacles. This type of cave allows the keener novices to explore crawls and squeezes, while easy bypasses are available for the less confident. It should permit one to two hours of activity, but at no time place the group more than about fifteen minutes from daylight.

The first caving trip is an important opportunity for education. The formation of caves can be explained, along with the importance of conservation, a subject that should be raised before the group encounters any vulnerable formations. Frequent short stops and explanations hold the attention much more effectively than one long, drawn-out lecture.

The pace of the trip should be relaxed but steady. Many dry caves are very peaceful places with a pleasant atmosphere. The group should sit down at least once and experience the total darkness of the environment. Instructions should be loud, clear and unambiguous. Asking the group to pass information back can work, but the 'Chinese whispers' effect often makes the advice useless by the time it reaches the last person.

Leading from the front is not always best. In some sections the group can safely find their own way while the leader follows on behind, or even takes a slightly different route. Any climbs where a slip can cause injury need to be protected by rope, or by hands-on supervision. This means being near enough to provide immediate physical support.

A successful trip should leave the group feeling well exercised but not exhausted. Ideally, they will have enjoyed the experience and the majority will be keen to repeat it.

Intermediate Groups

The novice caver has done a few simple trips and wants to progress to something more ambitious. This intermediate stage can be defined as involving trips that are likely to exceed three hours, where no short cuts back to the surface are possible, in caves that are strenuous and probably wet. Many club cavers choose to introduce novices directly to this level of difficulty when something easier might be preferable.

Swildon's Hole in Somerset illustrates this perfectly. It has an upper series of mainly dry, varied passages that interconnect and provide options suitable for novice groups. Trips exceeding two hours are possible without ever taking the group more than twenty minutes from the entrance. It also has a long streamway descending steeply with a short pitch, numerous waterfalls and pools ending at Sump 1. The trip to Sump 1 is frequently undertaken by novices under instruction who often find it quite exhausting. The upper series trip is more suitable, and yet is less often used.

One reason for this, and a recurring problem, is overestimating the abilities of the novice. Another is a blurring of objectives, where the needs of the novice become secondary to the sporting enjoyment of the leader. There is another odd attitude that emerges sometimes, a feeling among some cavers that novices are expected to be poorly equipped, cold, tired and miserable. That's just life if you're a novice!

At any level of instruction it is important to show the right way of doing things. For example, it is common practice for cavers to lifeline the novices down a short pitch and then to climb it themselves unlined. The message is clear – cavers do not use lifelines. Anyone introducing others to the sport has a responsibility to educate and to lead by example. It is all too easy for cavers to pass on bad habits and poor techniques. This is particularly important at the intermediate level as the adult novice caver is likely to progress quickly to being self-led, and opportunities to instil safe practice may be limited.

The club caving leader can have a difficult job managing a team of diverse individuals whose abilities may be unknown. Appearances and impressions can be deceptive: a caver may be well equipped and may have completed some reasonable trips, but may actually still be at a novice level. It is not a good idea to discover this, just after pulling a rope through at the beginning of a long, serious through-trip.

SRT

The SRT-proficient club caver often becomes involved in SRT training for club members or assists on club trips involving mixed-ability cavers. In my experience I have observed that the club SRT trainer actually has a more difficult job than the professional instructor. This is because the club trainer often has to deal with trainees who are at greatly varying skill and experience levels.

A lot of cavers buy the kit, learn the basics of SRT and then go for long periods without practice. The first sign of a problem is the person who cannot remember how to kit themselves up, or worse, doesn't realize that they don't know how to kit up! It really is important to have a practice session first and double-check up and down changeovers.

If in doubt, keep the rigging as simple as possible, and if necessary change the venue to something technically easy with short pitches. Rebelays are supposed to reduce time but this only applies to an experienced party and they are often the scene of long hung-up epics. It could be on this sort of trip that self-rescue skills are suddenly needed.

CHAPTER 11

Exploration

For the caver, the limestone hills underfoot are a puzzle within an enigma, shrouded by a mystery. In every region the caves that have been explored and mapped represent but a fraction of what remains dark and untrodden. Consider the Mendip Hills, the most intensively caved region in the south of Britain. Generations of cavers have dug on the surface at swallets and depressions and, underground, have dug, blasted, squeezed and pushed every possible passage. A hundred years of such activity have revealed,

Dolines in South Wales: there is very likely to be an unexplored cave somewhere beneath. (Chris Howes)

perhaps, 30km of passage. There is no way of knowing how much cave passage remains to be discovered in the area, but it is unlikely that more than 5 per cent of the total has been explored.

There is no greater thrill underground than to take the first footsteps through virgin cave passage, especially when the discovery results from long and hard work. Cave discovery in Britain is often very hard-won, which makes the big break, when it comes, all the more satisfying. There are no guarantees of success and vast numbers of man-hours have been invested in sites that are yet to yield results. Most club cavers put time and labour into digging but few actually experience the euphoria of original exploration.

In other regions of the world the situation is reversed. The high limestone plateaux of the Alps, for example, are riddled with an infinity of shafts, fissures and holes cunningly concealed in a chaotic lunar landscape that is buried under snow for nine months of the year. Most of the holes are choked with glacial debris less than 30m down, but some are the entrances to deep, vertical systems plunging down towards, and sometimes beyond, the magic 1,000m mark. An ordinary day's prospecting in this region means the routine descent of open and unexplored caves.

Further afield, throughout South East Asia for example, are vast regions of unprospected limestone. Every year expeditions visit areas in China, Vietnam, Sarawak and Papua New Guinea, where huge gaping entrances await investigation. We have barely begun the exploration of the world's caves.

Discovering New Caves and Passages in Britain

New cave systems are most often entered by digging on the surface. This is usually a deliberate act by cavers, but sometimes contractors break into cavities while excavating trenches or foundations. Many of these accidental discoveries go unrecorded, but sometimes cavers are invited to explore and survey, and occasionally a permanent access is possible.

Limestone quarrying frequently reveals caves and, sadly, usually destroys them as the working face advances. In Fairy Cave Quarry, Somerset, an extensive system of extremely well-decorated caves was broken into at several points. The quarry no longer operates, and the systems of Shatter Cave and Withyhill Cave remain intact, but the once-beautiful Balch Cave is a devastated ruin. The quarry is something of a poisoned chalice for the caver and our relationship with the industry is an uneasy one.

Stream sinks are often blocked by an accumulation of flood debris and sediments, while many other entrances have been buried by glacial debris. Some blockages are short and easily removed while others defy the intensive efforts of successive generations. Back in 1904 early Mendip cavers started work at Hillgrove Swallet and were confident of an early breakthrough; today, more than one hundred years later, work continues at the site and a major discovery is still awaited.

Any site that takes a reasonable stream or has an obvious draught is likely to reveal a cave. Depressions, or dolines, are usually formed by cavities underneath and are another good spot for excavation. It helps to have an understanding of the local geology when choosing a site, although caves are sometimes found at sites where, according to eminent experts, they should not really exist. 'Caves is where you find them', as one Mendip caver succinctly put it.

Passages often run virtually to the surface and are concealed only by a shallow deposit of topsoil. Recently, cavers exploring a new passage in Yorkshire reached a final blockage through which they heard a passing aircraft; a short spell of digging saw them squirming out onto the surface. The knowledge that many caves lie concealed has led cavers to experiment with location systems. One system is resistivity, which searches for cavities by passing an electric current through the ground, and another is the gravimeter, which looks for tiny anomalies in local gravity caused by underground cavities. Neither of these systems is entirely reliable or readily available to the ordinary caver. Ground-penetrating radar could be a more effective and practical tool for the future. This technology has now developed to a standard where it can penetrate tens of metres into

limestone. Affordable and user-friendly units are not available yet, but I expect cavers to be first in the queue when that day comes!

Warm air can rise from hidden caves during cold weather and vice versa. Snow may be melted by a warm draught, and cavers often go looking for obvious blowholes during the winter. This is a common way of finding caves in Alpine or Arctic regions. Sometimes vegetation is affected by the draught and the growth of certain ferns in a hollow or depression can be significant.

Cavers sometimes attempt to follow caves or find entrances by dowsing. This is a contentious issue that provokes much emotion among strictly scientific cavers. Several cave

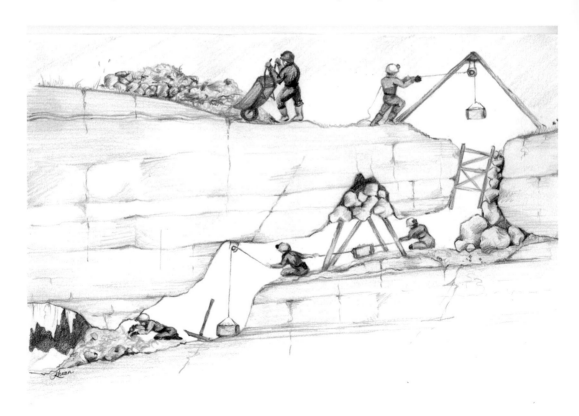

A cave dig – some are very labour intensive. (Rhian Hicks)

digs have been spurred on by extravagant predictions, but none has yet definitely proved the dowsers right.

Surface Digging

The first prerequisite for any surface dig is permission from the landowner. It is a good policy to fence off any excavation and always to leave it safely covered. Some surface preparation can be useful: access steps, for example, are best arranged before a depression gets muddy and slippery. Water diversion or damming may also be essential to allow an active swallet to be worked. Whatever you propose, always ensure that the landowner is consulted and informed as future access to any cave will depend upon good relations.

Most digs rely on muscle power, but with sufficient funds a mechanical excavator can be hired. This is particularly useful for large dry depressions where no specific point to dig is obvious. Several trial excavations can be made in the course of a day, until either an open or choked passage is located. An entrance shaft of concrete pipes is then usually installed before the depression is refilled and landscaped back to its original contours. There have been several successful digs using this approach in the Mendip area.

Most surface digs start vertically and then level off into a more horizontal passage. During excavation, temporary shoring using beams or scaffolding with wooden shuttering can be installed, but when the full vertical depth is reached a permanent entrance

Using a mechanical excavator, the easy way to get a dig started. (Chris Howes)

Templeton's Dig in the Mendip Hills, an amazing feat of engineering entirely excavating a large natural shaft more than 50m deep. (Elaine Johnson)

can be engineered. Scaffolding or steel drums are sometimes used to line a shaft, but concrete pipes, set onto a solid base, are the safest long-term solution. Much spoil is produced during this stage of a dig and this can be stored behind secure retaining walls. There is usually a good supply of rocks and stones from the dig that can be used for this purpose.

Spoil removal is often the most laborious and difficult aspect of digging. A winch and tripod can be set up over a surface shaft, but as the passage extends horizontally problems accumulate. Skips can be made by cutting a section from the side of a plastic drum. These can then be dragged along on their sides, or hauled vertically, while retaining (most of) their load. Sandbags made from hessian or woven plastic are another option that allow for removal and storage of spoil.

Muddy and arduous work, but potentially with huge rewards. (Rob Eavis)

Any serious dig must be kept to a good size. Excavating with a trowel at arm's length is only practical for very short obstructions, with a clear view into open space ahead. A passage about one metre square is a reasonable size and allows a workable face where tools can be used efficiently.

Explosives

If the passage is too small for forward progress, or blocked by large boulders, there is really only one practical solution – the use of explosives. In most regions there are cavers licensed to use 'bang' underground, and very few surface digs succeed without their help. Individual boulders can be shattered by a relatively small and well-placed charge, but for enlarging passages shot-holes are much more efficient. The cordless drill has become invaluable for this type of project and, used with extension batteries, can easily drill several holes of adequate length and diameter.

Fumes are a danger when explosives are used. Exposure can cause symptoms ranging from headache and nausea to, in extreme cases, unconsciousness and death. The speed with which fumes disperse depends upon the ventilation of the cave, governed by the direction and strength of any draughts. Some sites suck in very strongly, causing the fumes to vanish almost instantly, but others blow back towards the digging team, who may be forced to make a very rapid exit.

In digs without a draught, it may be necessary to leave the site for several days and it is important to place a warning notice, with dates, to prevent any visitors straying into the affected area. Fumes at their most dangerous are clearly visible as a white vapour. There is a strong marzipan smell, which lingers long after the visible vapour has dispersed.

Cave discovery in Britain is very dependent on the continuing use of explosives. It is essential that these explosives are used responsibly, by qualified persons, and always with the knowledge and approval of relevant landowners. Any concerns about usage in a particular cave should be addressed to the cavers concerned, rather than by reference to any licensing authority.

There are a couple of alternatives to high explosives that can be used. Snappers are small explosive charges that can be purchased and used without an explosives licence, although some training is required. They are really just giant 'bangers' with a built-in detonator fired using a cable and battery. Inserted into a shot-hole and well 'tamped' with mud, they will split a boulder quite neatly. The other method, which would certainly not be approved of by any health and safety officer, is called capping. This is a very effective way to shatter boulders, enlarge cave passages and lose the sight in one eye. I shall say no more.

Digging Underground

The ideal underground dig has a draughting airspace with a view through into an echoing void beyond, and can be made passable with no more than a few hours' work. Such sites are common in newly discovered caves and generally receive prompt attention. In more popular and established systems there is still great scope for digging, but the sites are liable to be much more long term. Any major passage ending in a constriction, mud or boulder choke may yield an extension, possibly a major one.

Understanding cave formation often helps to choose a good site and to avoid a pointless one. Cavers who do not appreciate the gravity-defying nature of phreatic passages, or the relationship between dip and strike control, are often at a disadvantage. Caves are unpredictable, and any dig can produce

the unexpected, but there are some poorly chosen sites that have consumed a huge amount of time and labour.

Digging can have a profound effect on the cave environment, and this must be a consideration before work begins. Sediments that have dried out slowly over centuries need only to be dumped under a drip to become a slimy porridge that is subsequently trailed throughout the cave. Spoil has to be dumped somewhere, and several cubic metres may be produced.

Think carefully about how to minimize the impact of any work. One well known Mendip digger begins by finding a water source, possibly just a drip, and arranges a header tank and hose pipe to keep the cave clean. Spoil is then carefully walled up into selected hollows and alcoves where it will have least effect. It is very common to take all manner of items down to aid digging – ropes, buckets, tools, hoses and so on – but there is often much less incentive to remove them after use. Many dig sites are not a good reflection on cavers' standards of conservation.

Passages often block with sediment at the bottom of a phreatic U-tube. The sediment level is usually the same at both sides and an equal amount of digging down and then up is required. This type of dig can emerge from under a calcite floor, and if a solid roof is met instead of an expected breakthrough it should be examined to see if it is crystalline. If it is calcite it can usually be broken by hammer blows, and the explorers emerge as if cracking out of an egg. There is often some water pooled above the calcite that can cause a fright to those below, but is unlikely to be of sufficient volume to cause a hazard.

The U-tube dig almost invariably floods, and often needs to be bailed out at the start of each session. The geology of Swildon's Hole has presented its explorers with many of these features. The famous 'Double

Troubles' appeared as mud chokes to the original explorers who excavated a route through. They quickly filled with water and now need to be bailed or siphoned to provide an adequate air space.

Carbon Dioxide Levels

Carbon dioxide is often a problem in cave digs. It is produced mainly by exhaled air, but can also be released by some disturbed sediments. Being heavier than air it collects at the lowest point, which is usually the dig face. If the air seems stale and workers quickly become breathless, it is almost certainly the cause. In a large passage, with easy access in and out of a small affected area, the hazard may be acceptable if each person spends only a short time at the dig face. But if the

An excavated shaft carefully shored to prevent collapse. (Tim Webber)

passage is tight or difficult it should be abandoned until conditions improve.

Boulder Chokes

Boulder chokes obstruct many major passages, particularly in the great cave systems of South Wales. They are highly dangerous obstacles that need to be approached with extreme caution. If the wrong boulder is disturbed, a major and fatal collapse is possible. Engineering is often required to progress safely and large quantities of scaffolding and concrete may need to be transported to the dig site.

The safest route is generally found by following a solid wall through the choke, assuming that one exists, or is heading in the right direction. There is often a cavity, or break-down chamber, above a boulder choke and this sometimes provides the easiest and safest route. Explosives have the major advantage over hammers and crowbars that they can be fired from a safe location. This is obviously crucial if removing a boulder that is likely to drop on to the digger below or might trigger a collapse.

The High-level Passage

Another way of discovering an unexplored passage is by climbing or using scaling techniques. Most caves have inlet systems that can be followed towards the surface, sometimes extending for great distances with a variety of interesting features. Exploration of inlets often ends at the base of a vertical shaft known as an aven. Some caves have extensive upper levels above the currently active system that can be accessed via avens or other passages leading off at high level. Any cave that is explored from the resurgence upwards can present its explorers with a succession of vertical obstacles.

There are several possible ways of climbing

Climbing using a cordless drill and expansion bolts. (Chris Jewel)

an aven. If it appears remotely free-climbable, a brave person (preferably an experienced rock-climber) can attempt an unaided ascent. The climber's safety is dependent on the correct use of nuts, pegs, camming devices and any handy natural anchors to provide 'running belays'. This technique needs skill and practice to be used effectively and should only be attempted by a climber with experience of leading routes above ground. Dynamic rope should be used. If the climbing is difficult and protection sparse, bolts will be required. A 12mm spit can be installed by hand in about twenty minutes, but if more than a couple are needed a cordless drill is the best option.

The more powerful models will help place several spits (and considerably more smaller bolts sometimes used for scaling) on a standard battery pack. It is important to drill the hole slightly short, and finish the last 2–3mm by hand. This ensures that the hole has a flat bottom and that the wedge will expand the anchor correctly.

In 2008 a team of British cavers established a world record for bolting when, during the course of a seven-day camp they reached a height of 213m in the remote extremities of Hirlatzhohle in Austria. The key to their success was a petrol engine drill, the fumes from which were able to disperse safely in the huge cavern. Even this mammoth effort did not succeed in reaching the top of the aven!

The usual method of bolt climbing uses stirrups or *etriers* to gain a little extra height between each placement. Efficiency can be further increased by using a bolting platform – a collapsible arrangement made from aluminium tubing that allows the caver to stand up, while supported behind the knees, and

Members of the team prepare for another scaling session. (Jeff Wade)

several sections of alloy scaffold tube (the length dictated by the preceding cave passage) clamped together, and then erected so that the end rests in, or against, a high-level passage. The maypole can be rigged with either ladder and lifeline, or with a single rope for SRT. The maximum limit for this technique is around 15m, or less if only short sections of tubing can be carried to the site.

The Virgin Cave

A well-known cave digger and speleologist has a philosophy that all digging cavers might be well advised to consider. The cave, he points out, is exposed to potential damage and destruction through their act of discovery. It is the explorers who cause this danger, and therefore their responsibility to ensure that adequate measures of conservation are applied.

Conservation

When a new cave or passage is entered, the team is elated and keen to investigate every possible option, but there are good reasons for moving cautiously. Virgin passages often have extremely nasty poised boulders that are liable to move with the lightest touch, so extra care is essential on slopes and especially where there are chockstones or boulder chokes. Conservation should begin on the exploratory trip, which means trying not to trample mud floors when a single path can be followed and being especially careful of vulnerable formations. Quite apart from safety and conservation, there is another good reason for moving slowly – you may have worked long and hard for this, so savour the moment.

A 'maypole' used to reach a high-level passage.

increases the distance between bolts to more than 1.5m. If this is still not enough, Australian cavers have even devised a climbing pole that extends 2 or 3m above the highest bolt. This is an extremely technical and advanced system, fraught with hazards, for which a very brave (stupid or gullible) caver is an essential requirement.

A simpler method of climbing shorter avens is by using a 'maypole'. This is usually

OPPOSITE: Looking up the immense aven in Hirlatzhohle during the record-breaking climb. (Jeff Wade)

Mud formations are very vulnerable to damage during exploration. (Brendan Marris)

Delicate crystal floors can be very difficult to preserve. (Rob Eavis)

Taping is the most immediate way to minimize damage and should be begun urgently to establish pathways and conservation areas. Remind photographers that tapes apply to them, too. You may decide that some vulnerable areas should be entered and photographed once only and then taped off permanently. In one Mendip cave an obvious enticing passage leads off between fine formations. The explorers have taped this off, but left a small survey on display to illustrate the extent of the passage and deter any future re-examination.

Restricted Access

The crux of the problem, and the cavers' dilemma, is that the ultimate protection for caves is restriction of entry. Attitudes to access controls are very regionalized. On the moors of Yorkshire the open cave or pothole is a feature of the landscape and the notion of having ugly gates or lids fitted is resisted fiercely. The sentiment that access to open ground should be free to all strikes a resonant chord for many northern cavers, who have little time for absentee landlords and permit systems. Gated caves and leadership systems are not usually a feature of caving around the Dales. In this area of the country, conservation relies entirely on the individual caver.

There is another way to preserve, and that is not to publish – or at least not to advertise – that an extension is well decorated. There are some cave surveys that, for conservation reasons, have never been released by their discoverers, but this approach can lead to resentment from other cavers who feel strongly that all such information should be in the public domain. Emphasizing hazards and difficulties can deter visitors, especially the inexperienced. Natural defences are often the best protection and it is not entirely unknown for discoverers to create their own in a bid to preserve. I know of one well-decorated cave where a short cut to the best grottoes was mysteriously blocked by a 'fallen' boulder.

In the southern regions it is now unusual for new caves to have unrestricted access. Another important factor here is the landowner, who will need to approve any access agreement. Increased concern about public liability discourages landowners from having open cave entrances, and gating is often at their request. Access is normally agreed through the club that has opened the cave, or through the regional caving council.

One of the simplest ways to conserve is to keep party size down to a minimum, and several caves in the Mendip area are restricted to groups of only three or four cavers. This also has the benefit of reducing nuisance on the surface and helps to maintain good relations with landowners. Novice cavers are banned from several sites because they are thought to have more environmental impact. While this is to a large extent true, a hundred parties of well-supervised novices may do less damage than one clumsy high-speed caver.

The most effective way to preserve a cave is by a system that only permits access to parties accompanied by an approved leader: approved in this sense means appointed by the controlling club or body, rather than

A cave surveyor takes a bearing using a hand-held compass. (Chris Howes)

holding an official BCA leadership award. One of the first caves to be controlled in this way was Saint Cuthbert's Swallet in the Mendip Hills. In 1953 the explorers found themselves in a highly complex system of passages, chambers and outstanding formations. They were determined to preserve their fabulous discovery and at once initiated a strict leadership system. Now, over fifty years later, the cave is extremely well preserved.

Some cavers will argue that this type of system is too restrictive and denies the caver much of the sporting satisfaction of finding and negotiating his own route. Some cavers feel this so strongly that they break open gates and 'liberate' locked caves, but generally it is accepted that conservation comes first, and that adequate open access caves exist for novice groups and casual cavers.

Surveying

Another important obligation for the explorer is to ensure that all new passages are surveyed to an appropriate standard. The British Cave Research Association has defined the grades of survey that might be made:

Grade 1 A sketch of low accuracy where no measurements have been made.
Grade 2 A sketch that is intermediate in accuracy between grade 1 and grade 3.
Grade 3 A rough magnetic survey. Horizontal and vertical angles measured to ±2.5 degrees, distances measured to ±50cm, station position error less than 50cm.
Grade 4 A sketch that is intermediate in accuracy between grade 3 and grade 5.
Grade 5 A magnetic survey. Horizontal and vertical angles measured to ±1 degrees, distances measured to ±10cm, station position error less than 10cm.

It is important to note that the grade of a survey represents its true accuracy, and not its intended accuracy. Not all surveys prove to be at the standard claimed. Working under the guidance of an experienced surveyor should ensure reliable results, but if no such expert is available it is wise to have a practice in a known site and compare results with an existing survey. Much has been published on the subject, but the best source of reference is probably Bryan Ellis's *An Introduction to Cave Surveying*, the second volume in the Cave Studies Series published by the British Cave Research Association.

Surveying Technique

The traditional instruments most often used are made by the Finnish company Suunto, which manufactures hand-held compasses and clinometers that are light, compact and robust. A tape measure made from PVC reinforced with glass fibre is recommended: the ideal length for British caves is 20m. A notebook is required to record the data, and specially made notepads with waterproof paper are available.

There is a magic little box called a Topofil that combines the functions of compass, clinometer and tape measure. It measures distance by unwinding a drum of cotton thread and displaying the distance on a counter, while an integrated compass and clinometer can be sighted along the thread to give bearings and inclination. This technique is not usually used in Britain, but its continental proponents claim speed, ease and accuracy.

A new generation of electronic devices is now superseding these instruments. There is the disto, which measures distance using a laser, and the curiously named Shetland Attack Pony, which combines an electronic compass and clinometer. These devices store data, which is useful but does not dispense with the need to make conventional notes and sketches.

Whether using old or new technology the basic procedure is the same. The surveyor must select a station, which is a physical feature such as a pointed rock or protrusion on the passage wall, and take measurements and bearings from there to another station. These are recorded along with details of passage height, width, shape and features between the stations. The procedure then continues from station to station through the cave. Some major caves have permanent survey stations to ensure accuracy, and these can sometimes be seen in the form of small triangles marked on the passage walls, or even metal studs hammered into drilled holes.

Recording data in the far reaches of Ogof Draenen. (Chris Howes)

When the data has been collected, the survey can be drawn. Most surveys consist of a plan and section. The section may be a projected or extended elevation. A projected elevation illustrates the cave as it would appear from one point in space and may have a foreshortening effect on some passages. An extended elevation effectively unfolds the cave into a straight line and presents it in a more meaningful way. Vertical caves are best represented by extended sections, especially if they have a spiral nature making a plan difficult to interpret; for horizontal caves the reverse is usually true.

It is advisable to calculate and plot the station positions rather than attempting to draw the survey directly from the data using a ruler and protractor. This is now almost routinely done using a computer, and much dedicated software is available: Survex is a popular program with many useful features. A huge benefit of digital surveys is the ability to view and rotate them in three dimensions, which can very effectively reveal relationships between different passages. With the main line of the cave plotted, the detail can be added using the conventional survey symbols.

Survey Discipline

When a new passage is discovered in Britain, initial exploration is often completed before thoughts turn to surveying. This approach may be appropriate for the rate of discovery here (except perhaps in the recently discovered Ogof Draenen), but much greater discipline is necessary on expeditions. The Americans have a motto, 'survey as you go!', which means that all exploratory teams survey, and move at the speed of survey. This requires great self-control when the cave has suddenly gone big, and huge passages beckon, but it recognizes that if the first team does not survey, it may never be done.

CHAPTER 12

Mines

Pick marks testify to hard labour beneath the moors of Derbyshire. (Rob Eavis)

The scale of some mine workings is colossal. (Brendan Marris)

Abandoned mines can be found in almost every county of Britain and provide a rich variety of historical interest, often combined with conventional caving difficulties. Those situated well away from the caving regions are often treated rather like substitute caves by local enthusiasts. For some enthusiasts mines, with their industrial archaeology and related social history, are actually of more interest than caves and their spare time is dedicated to exploring, surveying and attempting to reopen forgotten workings.

The history of mining in Britain goes back more than 4,000 years and begins with the extraction of flints from sites in southern England. The best-known is Grime's Graves in Norfolk where the prehistoric miners, using only antlers as picks, excavated deep pits with low galleries radiating at the bottom. The first mining for metals can be

Mines in limestone areas, like this one in the Forest of Dean, can contain calcite formations to rival those of natural caves. (Greg Jones)

More exclusive to mines are vivid blue formations associated with copper.

traced as far back as the Bronze Age, when short workings were extended underground to extract copper and tin. The arrival of the Romans, who were accomplished miners, saw the beginning of mining on an industrial scale. Their quest for lead, copper, iron and gold created extensive workings, some of which, like Roman Mine in Draethen, Gwent, can still be seen today.

The Roman expertise departed with them and mining in Britain declined, not fully recovering until after the medieval period. The early Industrial Revolution brought a surge in demand for coal, metals and other ores, creating a massive expansion in the mining industry. Stone was also quarried underground from sites around the country to supply the ever-expanding demands for building material.

Mining activity was on a such a huge scale, over such a long period, that the total number of sites is unknown. In this respect it can have the same fascination for the digger and explorer as the natural cave. There is also much scope for the researcher to study archive mining records, which can locate lost systems. When old workings are entered there are often historic relics such as tools or cranes to be seen, and inscriptions, perhaps centuries old, left by the miners themselves.

Mines can be a time capsule of industrial archaeology. (Brendan Marris)

Part of the extremely extensive stone quarries of Wiltshire where the famous Bath stone was excavated. (Brendan Marris)

Conservation and Hazards

The same rules of conservation apply to mines as to caves. Relics should be left in place, the walls should not be defaced and every effort should be made to preserve the environment as it was on the day the miners left the site. There are many hazards in old mines that are common to those in caves and some that are quite specific. The National Association of Mining History Organisations (NAMHO) has identified the following points:

- Never enter disused coal mines, which are prone to collapse and to bad air.
- Beware of, and do not handle, old explosives and detonators.
- Beware of rotten wooden floors that may be covered in mud, rocks or even water. There may be a deep drop below.
- Do not touch or disturb roof supports.
- Do not disturb, or climb on, piles of waste rock (known as 'deads').
- Hoppers were used to release rock into wagons; they may still contain much loose material held back by rotting timbers.
- Beware of flooded shafts in the floor, particularly in wading passages.
- Shafts are often loose and dangerous. They are sometimes partially boarded over with rotten timbers. Avoid touching the shaft sides, which are often only dry-stone walled ('ginged').
- Do not use wooden stemples, platforms or ladders as they are very likely to be unsafe.

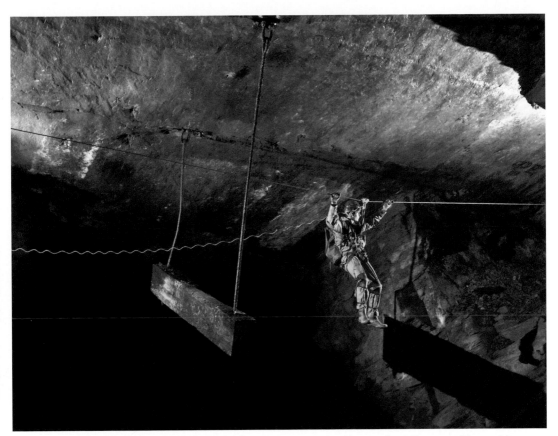

Mines can offer sporting obstacles like this wire traverse in the excellent Croesor–Rhosydd through-trip in North Wales. (Rob Eavis)

- Do not climb on old machinery as it may be weakened by rust.
- Beware of gas. Carbon dioxide accumulates at lower levels and will cause shortness of breath.
- Hydrogen sulphide can also be a danger and has a distinct bad eggs smell.

Route-marking

One hazard has been omitted from the list because it warrants special mention: the danger of getting hopelessly lost. Mines can be very extensive, with passage lengths running into tens of kilometres, and very often, particularly in stone mines, the workings cover an area several kilometres square. The pillar and stall method of mining can remove an entire bed of rock leaving only sufficient pillars to (possibly) prevent collapse. The result is a criss-crossing labyrinth of galleries with very few identifiable features.

This type of working can be seen in several locations including the extensive Box Mines in Wiltshire. This particular system has been well route-marked and, while the profusion of painted arrows and names may be unsightly, they have probably prevented several rescue searches. In systems that are not route-marked plastic arrows (flower pot labels are useful) can be used rather than defacing the walls.

CHAPTER 13

Cave Diving

The first generation of cavers was frequently frustrated when the cave roof met water and a sump barred the way. This was especially true of resurgence caves that often emitted huge volumes of water, indicating major unexplored systems beyond. It was in such a cave, Wookey Hole in Somerset, that the first successful cave dives were accomplished. This was in 1935, and still in the era of 'bottom walking'. Air was hand-pumped along a hose to the diver, who wore the traditional lead boots and brass helmet. With great difficulty the diver trudged through the deep sediments of the cave floor while dragging the hose along behind. There could hardly be more unsuitable equipment and yet the divers succeeded in entering new chambers.

During the same period diving attempts were being made at Swildon's Hole, a major feeder of the Wookey system. A sump (now known as Sump 1) had been reached fifteen years previously and had halted exploration of this very fine cave. The bulky conventional diving kit of the time was quite unsuitable for this site and home-made equipment was constructed. Luckily the sump was short and easily passed, allowing the explorers to proceed in a large streamway beyond, until stopped by a second sump.

It was in Wookey and Swildon's in 1946 that post-war self-contained equipment was first used. The divers were now free-swimming and very much more mobile. Major success came in 1948 with the discovery of the very big 9th Chamber in Wookey Hole. The first British cave-diving fatality was in Wookey the following year. The accident was caused by a faulty gauge and emphasized the dangers of cave diving and the limitations of the equipment then available.

The 1960s saw the introduction of the compressed air aqualung and diving suddenly became a more practical way to explore caves. There was quite intense activity with the new, more compact and reliable equipment around all the regions of Britain. Major discoveries included much of the Ogof Ffynnon Ddu and Little Neath River cave systems in Wales, extensions to Peak Cavern in Derbyshire and Ingleborough Cave in Yorkshire. Wookey Hole and Swildon's also yielded further major extensions.

Divers during this period had to learn and develop techniques suitable for longer and deeper dives. It became normal practice for divers to use two complete diving sets, enabling longer dives combined with higher security. The importance of the rule of 'thirds', which means always leaving one-

Cave diving is at the cutting edge of technical development. (Chris Howes)

third of the air supply for emergencies, was also recognized and is now a fundamental principle. Cave divers rely on a guideline to navigate through the often very murky conditions, and the importance of very precise line-laying became apparent as more complex sites were explored.

The techniques were well developed by 1979 when Geoff Yeadon and Oliver Statham completed the first through dive from Kingsdale Master cave to the resurgence at Keld Head, a world record distance of 1,840m. At around the same time, the final sump of Wookey Hole was being pushed by Martyn Farr to over 70m depth. There were, sadly, several fatalities during this period. The leading diver Roger Solari died following a fatal misjudgement at the end of a long dive, and there were other cases of cavers wrongly assuming they could simply strap on the gear and teach themselves.

The pinnacle of achievement in British cave diving came in 1991 with the connection of King Pot to Keld Head. The through dive, another world record at a staggering 3.05km, was completed in five hours by Geoff Crossley and Geoff Yeadon. The link also made the Keld Head system, at 6.5km, the longest underwater cave in Europe.

Triumph and tragedy in cave diving are never separated for long. In 1994 two leading divers, Ian Rolland and Sheck Exley, died in unrelated incidents in Mexican caves. Both were pushing the technical and human limits, Sheck Exley attempting (he may have succeeded) to break the 1,000ft (305m)

Squeezing through a constriction in Pwll y Cwm, South Wales. (Duncan Price)

depth record, and Ian Rolland using an extremely sophisticated rebreather system.

The rebreather has a huge advantage in gas economy over the conventional 'open circuit' aqualung. A breath inhaled from an open circuit system using ordinary compressed air is about 21 per cent oxygen. The diver absorbs only 25 per cent of this oxygen and the rest is simply breathed out as exhaust bubbles. The rebreather recycles the exhaled gases through a chemical scrubber, a carbon dioxide absorbent such as soda lime, which removes the carbon dioxide from the gas mixture and leaves the oxygen and other gases available for rebreathing.

Rebreathers allow much greater durations, particularly when diving at depth. The technology was instrumental in achieving a British cave diving depth record of –90m in Wookey Hole in 2005. Divers Rick Stanton and John Volanthen developed their own compact rebreathers for a series of record-breaking dives in this cave system, which has, since the inception of cave diving, been a proving ground for both divers and evolving technology. They were able not only to reach this considerable depth but to explore a constricted horizontal passage that would test most cave divers in shallower and less remote locations.

Most British diving is undertaken as a means to an end, and the objective is to reach explorable 'dry' passages. There are some sites where the passages are large, the visibility good and the experience of weightless caving can be enjoyed for its own sake. Wookey Hole, Keld Head and the nearby flooded systems of Chapel le Dale are the best-known British examples, but the very best cave diving is to be found across the Atlantic.

Rick Stanton about to begin the record-breaking dive in Wookey Hole. (Duncan Price)

In Florida, Mexico and the Bahamas there are extensive flooded systems that, not many thousands of years ago, were spacious air-filled caves adorned with spectacular formations. This type of environment is the cave divers' dream as they swim in crystal clear water among exquisite drowned grottoes.

These caves are often entered by flooded shafts such as the 'blue holes' of the Bahamas and the cenotes of Mexico. In Mexico, in 1996, the first dives were made into what has become known as Sistema Ox Bel Ha, now the world's longest underwater cave with a staggering 172km of explored passage.

It must be emphasized that cave diving is extremely hazardous if undertaken without appropriate training. Being both a caver and

LEFT: The glamour of British cave diving. (Rob Eavis)

an open water diver is not a sufficient qualification, as there are techniques and hazards
that are quite specific to the flooded cave.
Cavers who want to dive should join their
local section of the Cave Diving Group
(CDG), who can provide quality training.
The CDG has published a comprehensive
manual on the subject but stresses that this
is an aid to its training programme and definitely not a 'do it yourself' book.

LEFT: Stargate, a 'blue hole' on Andros Island
in the Bahamas, dramatically illustrates the
attractions of diving in warmer, clearer
water. (Chris Howes)

CHAPTER 14

Cave Photography

There are no spectators in caving and the only way we can share the richness and variety of our experiences is through the still or moving image. Most cavers have a go at photography underground in an attempt to capture and convey the pleasure and fascination that caves hold for them. It is a hostile environment for delicate equipment and a difficult medium to photograph successfully, but one where truly evocative results are possible.

It may be that you just want to take a few snaps with a disposable camera, in which case there are a couple of tips that are helpful. Remember that the effective range of the flash is only about 3m, so you need to keep the subject close. The other problem you will have photographing damp cavers is 'fog'. You may not see it but the camera will when the flash bounces back, creating a white mist on the picture. Avoid this by taking pictures before people get wet or by snapping as quickly as you can when damp people arrive at a suitable location.

The digital revolution has had a profound effect on the ease and quality of cave photography and few practitioners still use the medium of traditional film. If film is your preference the ideal camera will have a range of manual controls including a 'B' setting that enables the shutter to be left open.

Whether you choose film or digital, the basic principles of cave photography are the same and a built-in or camera-mounted flash is unlikely to give you an image of quality.

Flash and Lighting

Take a look at the four examples of the same view shown here and compare the first picture, taken with the camera's own flash, to the others. Apart from being rather 'flat', it has mysterious 'orbs' caused by the light reflecting from suspended water droplets.

Compare this with the second picture taken by using a separate flash fired from just one metre to the side of the camera and you will see the image is more dramatic and the 'orbs' have gone. Fog will also be avoided by moving the flashgun away from the camera.

In the third example a flash fired behind the subject has created depth and highlighted the natural architecture of the passage. This back-lighting is a key technique in producing dramatic pictures.

The water in the first three images is frozen in movement by the speed of the electronic flash, but in the fourth image it is fluid, because no flash has been used. The camera has been mounted on a tripod and the subject has been illuminated by caving

Simple photography using the camera's built-in flash.

A tripod-mounted time exposure using caving lights alone.

Adding character with a separate flash.

Using back-lighting to create a dramatic effect.

lights during a two second exposure. The LED lighting has a different colour temperature that creates a bluish tone to the picture and an altogether different 'mood'.

The easiest way to experiment with creative lighting is to set the camera to a long exposure (a couple of seconds) and then fire a flashgun when the shutter is open – a technique known as 'open-flash'. One problem with this is that it has to be done in the dark, which the autofocus on a modern camera may not like. This is where having full manual control really helps.

If you have more than one flashgun there are a couple of ways you can utilize them. One method is a countdown where the subject stays as still as they can (in complete darkness) and the flashes are fired as simultaneously as possible. Even a tiny amount of camera shake between flashes will spoil the image, so a tripod or convenient flat surface is an essential for this technique.

Another option, which does not require a tripod, is to use slave units, electronic triggers that fire one flash in response to another. Using the camera's own flash to trigger a slave is a favourite system that avoids complete darkness and the possibility of the subject or camera moving between flashes. More advanced slaves have the option of an

Flashbulbs are much sought after by cave photographers because of the effects they can create with water. (Chris Howes)

A striking effect created by projecting a shadow onto a waterfall using a powerful spotlight. (Rob Eavis)

Electronic flash freezes the movement of water. (Chris Howes)

infrared trigger. If you do not want to use the illumination from the camera's flashgun you can tape an infrared filter to the window to provide an infrared only flash. The slave unit will be triggered, but no visible light will come from the trigger flashgun on the camera.

There is a major problem using this technique with many digital cameras: the double flash. Most modern cameras fire the flash twice; the initial pulse is for focusing (which fires the slave) before the main flash fires to illuminate the picture. Help for this problem is at hand with the ingenious Firefly 3 slave unit, which can be programmed to ignore pre-flashes, and has the added benefit of an infrared trigger.

The 'open-flash' technique can be used to illuminate a very large cavity with only one flashgun. Traditionally this was done by using a tripod and leaving the camera shutter open while the flash operator moved from one location to another using minimal light so as not to leave a trail on the film. This works fine with film but not with digital cameras because prolonged exposure creates 'noise' on the image and greatly reduces the picture quality.

This is where other digital techniques come to the rescue and give a huge advantage. With the camera solidly fixed on a tripod, the photographer can take multiple shots to be combined later using a computer program like Adobe Photoshop. There is no danger with this technique that one mistake will ruin the shot – in fact you can fire flashguns all over the place in any direction you choose! This is because you can select only the details you want from each shot – maybe just one small highlight – to add a subtle touch here and there.

It's a fascinating creative process that allows quite different interpretations from the same source material. The technique can be taken a stage further by panning or tilting the camera to cover a wider area. The accompanying shot of Golgotha Rift has been compiled from more than thirty separate exposures, including camera angles up to 45 degrees both up and down. The result is an 'impossible' picture that very effectively illustrates what conventional photography could not. It should be said that the picture took only about an hour to shoot in the cave but more than twenty hours to create on the computer!

There are many more tools and effects you can use on the computer, from simple brightness and contrast adjustments to advanced techniques like cloning.

Video

With digital video cameras becoming smaller, cheaper and more light-sensitive, recording moving images underground has become relatively easy. For really creative illumination you need a couple of lamps around 100W each, one of which can be used for dramatic back-lighting. A favourite effect is to start with the back-light as cavers approach the camera and then to gradually bring in fill-light to illuminate them. It can look superb but often needs quite a few rehearsals to get all the timing and movement right.

I have been making films underground for more than twenty years and it has been fascinating to witness the changing technology. The production of films like *Solo* and *A Rock and a Hard Place* required enormous amounts of heavy equipment to be transported underground. What a contrast that is

Golgotha Rift, Reservoir Hole, Mendip Hills: an 'impossible' picture assembled from multiple images. (Andy Sparrow)

with my recent film *Matienzo Diary*, which was shot entirely by the lighting of cavers and filmed on an ultra-lightweight camera. A film shot under these conditions has technical limitations but it can very effectively capture the spirit of the moment – especially during original exploration.

There are some general principles to consider when shooting, especially if you have no access to editing facilities. Firstly, keep all the action moving consistently from right to left or vice versa, and only change this if you wish to indicate that the party has reversed direction in the cave. Do not cut from a subject and then start to video again from a similar position or you will have a jump-cut. Cut away to another subject (someone watching or some stalactites) or to a big close-up of the subject's hands or face from a different angle.

The same general rules of lighting apply as for still photography. Experiment, in particular, with back-lighting to highlight passage features. If the camera has a live monitor you have one huge advantage over the cave photographer, which is the ability to see the effectiveness of your lighting arrangements through the camera. Move the camera smoothly and slowly (imagine you are immersed in a very thick liquid) and only use the zoom (again, slowly and smoothly) when it serves a purpose.

Art

The underground landscape has inspired many cavers to sketch, paint or even sculpt. In 1994 the International Society for Speleological Art (ISSA) was formed by a group of cavers and artists who shared a common interest in cave-related art. The ISSA organizes caving meets and exhibitions around the country. More information and examples of their work can be seen at www.issa.org.uk.

Robin Gray: *The Ladder Climb*.

Robin Gray: '*Up, Up, Up*'.

CHAPTER 15

Where to Cave

Discovery and exploration continue throughout the world and every year extensive new systems are explored. At the time of writing, there are eighty-nine caves over 1,000m deep, more than 500 caves exceeding 3km length and 742 more than 5km. Even within intensively explored areas new discoveries continue, and there are entire caving regions around the world that have barely been examined.

United Kingdom

The major caving areas of Britain are restricted to outcrops of mainly Carboniferous and Devonian limestone (shown as blue on geological maps). The younger, and very extensive, Jurassic limestones seem to have only limited cave development, the only notable exception being the Isle of Portland, where the tortuous system of Sandy Hole has been extended to well over 1km.

Each caving region has a unique geology and landscape, and offers a different type of caving. The systems of Yorkshire are often vertical and very active, while the caves of South Wales are horizontal and extend for great distances. Both have their own hazards, make their own demands on the caver and provide quality caving experiences. The caves of Mendip, the Forest of Dean and Devon have their own fascination and offer some very challenging caving, often amid the finest of formations. The British caver who seeks the best that every region can offer will experience a wealth of varied challenges and rich subterranean landscapes.

Devon

There are no active stream caves, but there are extensive old phreatic systems, originally formed by the River Dart on the fringes of Dartmoor. The most popular of these are Pridhamsleigh and Baker's Pit; both are extensive and very complex. 'Prid' is famous for its mud and also a deep lake that has been passed by divers to reveal the biggest chamber in Devon. Afton Red Rift is another important system that provides a round trip including some tricky traversing. There are many minor caves and much unfulfilled potential to inspire the cave digger.

Formations in Afton Red Rift, Devon.

Mendip

At the start of the twentieth century only a small number of caves were known in the Mendip Hills of Somerset. Apart from Wookey Hole and Cheddar (where conducted tours were already running), there was Goatchurch Cavern at Burrington, which is still one of Britain's most popular beginners' caves. Lamb Leer, a cave with a huge dome-shaped chamber, originally discovered by lead miners in the seventeenth century, was one of the few other caves known during this period.

The first generation of Mendip cavers found one great system open and awaiting their attention – Swildon's Hole. From its little entrance near Priddy church it led them

The magnificent Main Chamber of GB Cave. (Andy Sparrow)

Traversing the 'Double Pots' in Swildon's Hole. (Brendan Marris)

down into spacious and beautifully decorated galleries. Each subsequent generation of Mendip cavers has added to the system and it now extends for more than 10km.

Other swallets in the region were not so accommodating, however, and were almost without exception totally choked. The pioneers realized they would have to dig to find caves and so began a tradition that continues unabated today. The first success came in 1904 with the discovery of the deep, strenuous system of Eastwater. This is classic Mendip caving, the development being dominated by the steeply inclined bedding planes. The cave descends in a giant spiral and, as many a caver has found out, is much more tiring on the upward journey. The very severe West End Series and the notoriously difficult Primrose Pot provide serious challenges for the slimmer caver.

GB Cave, discovered in 1939, is in total

contrast: an easy, impressive system with a massive chamber and fabulous formations. Soon after this, Stoke Lane Slocker on the eastern Mendips was pushed through a short sump to reveal a series of large decorated chambers, and Longwood Swallet was excavated to reveal a sporting and extensive system beyond tight entrance squeezes. Several shorter caves have been opened at Burrington, most of which are suitable for beginners.

Most of the remaining major swallets have now given up their secrets. The great system of Saint Cuthbert's revealed a vast complex of chambers, passages and outstanding formations. Manor Farm Swallet offers a steeply descending decorated stream passage leading down to intercept a section of much

Admiring formations in Shatter Cave. (Andy Sparrow)

The sporting streamway of Longwood Swallet. (Brendan Marris)

larger fossil passage, and Upper Flood Swallet has recently been extended to reveal a major system over 3km long. Two digs have entered caves that are unusual for Mendip, being mainly vertical in nature. Rhino Rift is a huge natural shaft draped in flowstone, while Thrupe Lane Swallet is an active system with a variety of pitches, one option being a spectacular 60m shaft.

The local quarrying industry has also revealed, and often destroyed, some extensive and extremely well-decorated systems in the eastern Mendips. Withyhill Cave and Shatter Cave are largely intact and have many excellent formations. The main resurgences at Cheddar and Wookey have both had major extensions through diving. Both caves currently end in very deep sumps requiring highly complex decompression profiles.

Recent finds have included the main streamway of Wigmore Swallet, which has now been dived through nine sumps to reach a boulder choke. The stream has been tested to Cheddar Risings, more than 9km away, with a depth potential of more than 300m. In 1996, intensive work by the Bristol Exploration Club forced a connection from the tight and difficult Priddy Green Sink to Swildon's Hole, creating a superb and varied through-trip.

There are many minor caves in the area offering short but interesting trips and exploration continues with unabated enthusiasm. There are several club hostels around Priddy, and most cavers frequent the Hunter's Lodge nearby.

Forest of Dean

The River Wye, flowing south from Monmouth to join the Severn at Chepstow, has incised a deep valley into the local limestone. Between the two rivers is the Forest of Dean, an expansive area of wooded hills with a long history of coal and iron ore mining. The many stream sinks in the area and the tendency of the extensive iron mines to intercept sections of ancient natural cave encouraged much digging from local cavers.

For many years the natural caves were elusive but the first major discovery in 1974 fully showed the huge potential of the region. Otter Hole, situated on the bank of the Wye near Chepstow, was entered following a relatively short dig at a large entrance. The explorers followed a passage through glutinous mud crawls, an unusual tidal sump and finally into the most spectacularly decorated passages in Britain.

Keyhole passage in Slaughter Stream Cave. (Brendan Marris)

The next major discovery came in 1990 when a dig at Wet Sinks, originally begun in the 1950s, broke through into the excellent Slaughter Stream Cave, a system of active and fossil passages more than 14km in length. The nearby Redhouse Lane Swallet yielded a major passage a year later and a link between the two systems is a strong possibility. Encouraged by these successes, digging continues at several promising new sites and 'The Forest' is rapidly developing as a major caving area.

South Wales

The limestone of South Wales extends from Abergavenny westwards, in an almost uninterrupted band for over 50km. Much of the limestone has a capping of Millstone Grit, making direct entry into the caves difficult, and most entrances are in the valley or escarpment sides. The limestone has only a slight dip and the caves are predominantly horizontal. The conditions are ideal for the formation of very long cave systems, many of which still await discovery.

Several caves have long been known, most notably Porth yr Ogof at Ystrafellte, where the River Mellte flows for a short distance through a spectacularly large passage. In the adjacent valley near Pen y Cae is the large

Beyond the Long Crawl in Dan yr Ogof. (Brendan Marris)

resurgence cave of Dan yr Ogof. Spectacular finds were made here in 1966 when caver Eileen Davis pushed through the notorious Long Crawl to discover a fabulous series of passages and formations. Some of the finest examples of straw stalactites and huge phreatic 'railway tunnels' can be seen here.

On the opposite side of the valley is the great cave of Ogof Ffynnon Ddu. The South Wales Caving Club dug into the lower end of this system, near the resurgence, in 1946. They followed a superb active streamway to a terminal sump and explored a complex high-level network. But this was only a fragment of the system and in 1966 a second major breakthrough revealed a vast complex of passages stretching up towards the major stream sink. Subsequent work opened two entrances at higher levels, creating some of the best through-trips in Britain.

During the same period, divers investigated the terminal sump of the nearby Bridge Cave, which conducted water from the Little

Neath River in a short but very large passage. The sump was easily passed and beyond were the great halls and galleries of the Little Neath River Cave. A side passage later provided a second and very wet entrance to the system, creating one of the most popular caving trips in Britain.

The major discoveries in recent years have been in the Llangatock area at the eastern end of the limestone. The first big find here was Agen Allwedd or 'Aggie', first entered in 1946 through a small entrance in the escarpment. The passing of a boulder choke in 1957 revealed huge fossil passages and a long active streamway. Further discoveries enabled some very long round trips to be made (including the classic 'Grand Circle') and confirmed the massive potential of the area.

An intensive dig in the Clydach Gorge in 1976 revealed the long, big and beautiful Craig y Ffynnon, representing the downstream end of the fossil system. The huge

The notorious entrance crawl of Ogof Daren Cilau. (Rob Eavis)

The streamway of Ogof Ffynnon Ddu. (Chris Howes)

gap between here and 'Aggie' was filled in 1984 by the cave Ogof Daren Cilau. This was already known as a long and arduous crawl leading to some big, but not extensive, passages. A boulder choke was passed to reveal a huge system of passages, some of unprecedented size. The end of the cave terminated in a sump that was passed by veteran cave diver Martyn Farr in 1985 to emerge after 600m in Elm Hole, a short cave near the resurgence in the Clydach Gorge.

There were many excellent dig sites deep into the cave and underground camping became established as the only practical way of working these. More extensions followed and the end of the cave became progressively more remote until trips of twelve hours were needed just to reach it. The gap

between Daren and Aggie has reduced to less than 30m, but despite dedicated and exhaustive digging at both ends the link remains frustratingly elusive. In the other direction the distance to Craig y Ffynnon has been reduced to a similar distance, but again the connection proves difficult.

Across the Clydach Gorge lies another major limestone outcrop running from Gilwern Hill south to Pontypool. No significant caves were known here until members of the Morganngwg Caving Club dug at the end of the short but strongly draughting Ogof Draenen. A break-through in 1994 led into a complex of large active and fossil passages. A bonanza of discovery has followed and the system is estimated now to be a contender for Britain's longest cave, with more than 65km of surveyed passage. There are numerous excellent trips in this cave,

The 'Blue Greenies' in the remote extremities of Ogof Daren Cilau. (Rob Eavis)

including an extremely varied and sporting 4km round trip.

North Wales

There are extensive limestone outcrops in North Wales and a small number of important caves. These are generally phreatic in nature and have been artificially drained by mining operations in the region. Ogof Hesp Alyn is more than 2km long, and descends a series of pitches that can flood from the bottom upwards to fill the whole cave. On Minera Mountain there are three major caves that probably represent sections of one very important system. Ogof Llyn Parc is the biggest and is entered by an old mine shaft over 100m deep. Access in the area is sensitive and cavers are requested to contact the North Wales Caving Club in advance of any trip.

Part of Ogof Draenen. (Brendan Marris)

Derbyshire

The main caving area of Derbyshire is in the Castleton region of the Peak District. This was extensively mined, mainly for lead, over a period of several centuries. Since the caves and mineral deposits formed along similar geological features, many systems were entered and modified by the miners. The Derbyshire caver consequently benefits from an understanding of mining history and potential mine hazards.

Peak Cavern is the most famous cave in the area and boasts one of the largest entrances in Britain. Beyond the showcave are extensive passages including some fine formations in the high-level White River Series. A long crawl connects with Speedwell Cavern, another showcave, where visitors (transported by barge along a mined passage) stop short of the superb active

The magnificent Titan Shaft. (Rob Eavis)

streamway and towering avens. Another entrance to this system is James Hall Shaft, a mined entrance leading down to some impressive natural shafts. The most recent entrance to the Peak/Speedwell system leads to the famous Titan Shaft. In 1999 an upward excavation through a boulder choke revealed a towering shaft of unknown height. Sixty metres of bolting upwards took the explorers to The Event Horizon, where the cavity expanded into a huge chamber with the roof still out of sight. Titan eventually 'topped-out' at 145m to become the greatest vertical cavity in British caving and a strong contender for the largest chamber. A surface dig was connected to Titan in 2003 creating a spectacular new entrance to this rapidly expanding system.

An important feeder to this system is Giant's Hole, a deep, active and sporting cave that connects through a long and tight crawl to the nearby Oxlow Caverns to give a total depth of 195m. Oxlow is a series of short pitches, but an alternative entrance, Maskhill Mine, gives much deeper drops (some are natural shafts), made hazardous by much loose rock. Close by is P8, a short but popular cave with a pleasant streamway and many formations. There are also two mainly vertical caves: Eldon Hole, an impressive open shaft 60m deep, and Nettle Pot. Nettle Pot begins narrow but enlarges below the 49m entrance shaft. A short pitch leads to the magnificent Elizabeth Shaft, a free-hanging 52m drop.

There are many other caves in the area including the very difficult Dr Jackson's Cave, with its awkward passages and dangerous traverses. Winnats Head Cave has a large chamber, but extensions at lower levels are dangerously unstable.

Further south near Monyash is Lathkill Head Cave, which can only safely be explored in very dry weather. A new entrance has bypassed exceptionally long crawls and

The entrance to Giant's Hole, one of the most popular systems in Derbyshire. (Rob Eavis)

allowed exploration to push onwards. A dry-weather connection has also been made with the nearby Ricklow Cave.

There are two mainly horizontal systems popular for beginners' groups: Carlswark Cavern at Stony Middleton, and Bagshawe Cavern near Bradwell. Bagshawe operates as an 'Adventure Cave' for novice cavers and is a handy place to arrange a first caving trip.

Yorkshire Dales

The area around the Three Peaks of Ingleborough, Whernside and Pen y Ghent is the most cave-rich in Britain and offers not only staggering quantity but exceptional quality. This is classic limestone country illustrating a profusion of karst features. The hills are composed of near-horizontal limestone capped with ridges of Millstone Grit.

Rainfall is high in the area and the catchments are often steep and extensive. The result is active and clean-washed caves that are very flood prone.

The largest system is near the western end of the outcrop beneath the isolated and expansive moor of Casterton Fell. More than fifteen separate entrances lead into the Lancaster Easegill System, which has a total

Tatham Wife Hole, a typical vertical system of the Yorkshire Dales. (Tom Philips)

passage length in excess of 60km. There is great scope for through-trips, the most outstanding of which must be the longest – the traverse from Top Sink to Pippikin Pot. This is almost the ultimate caving trip in terms of that most essential ingredient – variety. There are roaring streamways, ascents and descents of great gun-barrel shafts, huge fossil galleries, abundant formations, traverses, crawls, climbs and some difficult squeezes. Add to all this the complications of route finding through Britain's most complex system and you can expect a very full day's caving.

Lancaster Easegill is only a part of the potential Three Counties System. It has already been connected by divers to the splendid Lost John's Cave, a fine system of wet and dry pitches leading down to the impressive Master Cave. Another link, to the

nearby Notts Pot/Ireby Cavern System, is being sought by cavers above and below water. The system of Rift Pot and Lost Pot is another essential piece in the jigsaw, representing an underground watershed as separate streams flow in opposite directions: one into Ireby Cavern, the other into the West Kingsdale System.

Kingsdale offers some excellent caving including the superb pull-through trips of Swinsto Hole and Simpson's Pot. After a series of wet pitches these caves combine and lead through to the West Kingsdale Master Cave. This impressive passage with its powerful stream soon sumps, but a fossil passage leads on at a higher level and follows a classic phreatic tube to emerge from a small entrance near the valley floor.

The resurgence for the area is Keld Head, which has been the scene of several record-

Part of the extremely extensive Easegill Caverns. (Tom Philips)

Easy progress in the Kingsdale Master Cave.
(Rob Eavis)

breaking dives. Over 6km of flooded passage have been explored. Recent work has connected King Pot on the eastern side of the valley, while in the other direction the link with the Rift Pot and Large Pot area is being pursued. A connection is also possible with Dale Barn Cave. This has recently been entered from a dig in Kingsdale, but its original entrance is the next valley over, Chapel le Dale.

Chapel le Dale is dominated by the great escarpment of Ingleborough's western slopes. The slope is divided by a terrace of moorland containing several deep pots. There is Tatham Wife Hole, with its series of short pitches, Black Shiver, featuring a magnificent 75m shaft, and the excellent Meregill Hole. Meregill is located at the base of a huge natural funnel and is extremely flood prone.

On the north-eastern side of Ingleborough is the great open shaft of Alum Pot. This is more than 60m deep but can be entered partway down via the popular cave of Long Churn (vast hordes of novices permitting). Another route to the bottom is the exceptionally wet and hazardous Diccan Pot; one of the coldest and most hostile caves in Britain. Nick Pot on the moor nearby has an underground pitch over 100m deep, and the final pitch of Juniper Gulf, although shorter, is tremendously impressive.

The most famous cave on the hill is Gaping Ghyll. The river of Fell Beck tumbles over the edge of a 100m shaft and crashes down into what may be Britain's largest known chamber. There are several routes into the cave, all involving multiple pitches. Exchange trips between entrances are popular and, for the diving caver, the connection with the resurgence system of Ingleborough Cave offers a fine and serious through-trip. Twice a year a winch is set up over the main shaft by local caving clubs allowing easy access to the Main Chamber.

At Ribblehead and Birkwith, between Ingleborough and Pen y Ghent, there are several good beginners' systems offering short through-trips. Calf Holes, Birkwith Cave and Old Ing Cave are all recommended as excellent first trips, but like most caves in the region they are floodable.

Pen y Ghent Pot is situated below the summit ridge and offers a tremendous descent of twelve wet pitches. Recent discoveries have revealed 'The Highway to Hell', an appropriately named extension with much crawling. Nearby are the famous open shafts of Hunt Pot and the massive Hull Pot. On the eastern side of the mountain are the deep, mainly vertical systems of Dale Head Pot and Gingling Hole.

There are a few scattered caves above the

OPPOSITE: The Main Chamber of Gaping Ghyll.

famous Malham Cove but none has yet entered the massive system that must exist here. The resurgence below the Cove has been the scene of some determined diving and underwater digging but has not yet produced significant dry passage.

The area around Kettlewell and Grassington has many important systems. To the north are Langstroth Pot, Pasture Gill Pot and the notoriously tight but well-decorated Strans Gill Pot. A new system, Hagg Gill Pot, is an easier option and also has many fine formations. Further south, in Wharfedale, is the long through-trip of Providence Pot to Dow Cave. There are no pitches but much difficult traversing makes the going far from easy.

Two systems on the moor above Grassington have enormous potential, coupled with much drama and tragedy. Langcliffe Pot is a long and serious trip including some arduous passages and massive unstable boulder chokes. There have been rescues here caused by both collapse and flooding. The nearby system of Mossdale Caverns was the scene of Britain's worst caving accident in 1967 when six leading cavers drowned during a flood. The system remains an enigma; its extensive crawling passages perch more than 170m above the resurgence in the valley below.

Scotland

Scotland, though devoid of vast areas of limestone, has a few relatively short but particularly inspiring cave systems set in spectacular scenery and generally well away from the crowds. The Isle of Skye boasts a series of highly flood-prone stream caves including the 376m long through-trip Uamh Cinn Ghlinn – a real collector's item – and the attractive Uamh an Ard Achadh, both located a few miles south of Broadfield.

Appin, near Glencoe, is home to a variety of small and generally difficult-to-find sys-

tems in the Glen Stockdale and Glen Creran areas – Uamh nan Claig-ionn being one of the most sporting and deepest at 48m. It has some fine rock scenery.

The Mecca of Scottish caving is Assynt, Sutherland, where the Grampian Speleological Group has its field centre. The 1.5km-long Allt nan Uamh Stream Cave provides interest for beginners, diggers and cave divers alike, and may one day be connected with the 3km Uamh an Claonaite. This is Scotland's longest and finest system with a superb streamway, extensive dry series, huge chamber and selection of sumps. It is a cave of national standard and importance. A new entrance to Claonaite, Rana Hole, has recently been opened by digging and provides a dry route to Scotland's largest chamber, The Great Northern Time Machine, previously only accessible to divers. In the adjacent Traligill valley the Cnocnan Uamh system provides excellent wet caving with attractive formations and plenty of scope for extending its present 2km length.

Schiehallion (Perthshire), Durness (Sutherland) and many other minor limestone areas scattered throughout the country are also well worth visiting, bearing in mind the sometimes atrocious weather and the fearsome bane of the Highland midge!

Ireland

The limestones of Ireland are very extensive and relatively undisturbed by uplift and faulting. Due to these factors the caves are typically long streamways with a gentle gradient, and the surface exposures feature very fine rock pavements. The best examples are in County Clare, in and around the Burren, which is a classic example of limestone scenery. Pollnagollum is over 12km long and has some very attractive passages, clean-washed by frequent flooding. Other major systems in the area include the Cullaun

Caves, the Doolin Cave, more than 10km long, the flood-prone Collagh River Cave and Pol-an-Ionian, which boasts a 7m-long stalactite.

Northern Ireland has an important caving region in Fermanagh. Reyfad Pot has entrance pitches dropping 100m into fossil passages, which lead in turn to an active streamway. The system is over 6km long and descends to –179m. Noon's Hole begins with an 80m surface shaft that enters the streamway of Arch Cave to give a system over 4.5km long. Another impressive system is Marble Arch Cave, which has now been commercialized and transports visitors by boat through one section.

Continental Europe

Austria

Austria combines mountainous alpine landscapes with vast areas of limestone to provide some of the deepest and hardest caving in Europe. The high *lapiaz* plateaux of the Dachstein, Tennengebirge and Totesgebirge cover hundreds of square kilometres and conceal an unknowable number of entrances. Every year new deep systems are explored through a succession of tight jagged passages and huge shafts.

One system, the Lamprechtsofen, was explored from the bottom up to a height of more than 1,000m before being linked to the Verlorener-Weg-Schacht to create the world's deepest through-trip at 1,485m. This has been described as 'a nerve-shaking ordeal through some of the world's most unpleasant cave passages'. The system currently has a depth of 1,632m and ranks as the third deepest cave in the world.

There are some very extensive mainly horizontal systems including the 96km

Elaborate fixed aids in Hirlatzhöhle.
(Chris Jewel)

Hirlatzhöhle, the 63km Dachstein-Mammuthöhle and the Jägerbrunntrog system, which not only extends to more than 30km, but descends to over 1,000m. Ice is often a feature of the high-altitude caves and requires very specialized techniques. Two famous showcaves, the Dachsteineishöhle and the Eisriesenwelt, contain outstanding ice formations and are highly recommended.

Belgium

There are many minor systems in the wooded hills of the Ardennes, the best-known being the Han-sur-Lesse river cave, which operates as a showcave. There are few vertical systems, the Trou Bernard being the deepest at a modest –140m. Caving is very popular in the region, which has led to conservation and access problems similar to those experienced in the south of England.

France

Nearly 30 per cent of France is limestone country and over 20,000 caves are known. It is the country where caving originated, as both a science and a sport, and the land from which many pioneering cavers emerged. For several decades one world depth record

Taking a break in the Gouffre Berger. (Andy Morse)

discovery in France displaced another and it is only in recent years that exploration in other countries has stolen that long-held crown. The deepest cave in France is currently the Gouffre Mirolda at –1,626m, the world's first mile-deep cave, but currently ranked at number four in the list of the deepest.

Another former record-holder is the Jean Bernard at –1,602m, one of a number of deep caves in the Jura region near Mont Blanc. There are other extensive limestone areas extending southwards including the Parmelan Plateaux, which contains the Reseau de la Diau, a long resurgence cave that can be entered at the upstream end through the Tanne de Trois Betas to give a superb 607m deep through-trip.

Further south is the Chartreuse region and the Dent de Crolle system, which was immortalized in Pierre Chevalier's classic book *Subterranean Climbers*. The system is extremely complex and contains over 55km of passage. The six entrances allow many possible through- and exchange trips, which are very popular with visiting groups. This system is just one of 600 caves within the Chartreuse region.

Still further south is the Vercors, another extensive area rich in varied systems and a very popular venue for club trips. One cave here is renowned above all others in Europe, the classic of classics, the Gouffre Berger. In 1956 this became the first cave in the world to exceed 1,000m in depth. It has now been surpassed by a score of other systems and, while it can no longer claim to be caving's Everest, it has, perhaps, become a caver's Matterhorn. Its enormous passages and huge formations provide a stunning backdrop as the 'Starless River' is followed down a variety of impressive pitches to the terminal sump. The cave is extremely popular and usually needs to be booked at least a year in advance. It is well within the capabilities of the ordinary caver and many major British clubs arrange a visit at some time.

To the east, much limestone surrounds the

Massif Central, and the regions of Dordogne and Ardèche are well known for their spectacularly beautiful caves. There are some long horizontal systems and the showcaves of Padirac, L'Aven d'Orgnac and Aven Armand are well worth visiting. Major systems in the area include the 24km-plus Grotte de Saint-Michel d'Ardeche and the 22km-plus Reseau de Foussoubie.

South lie the Pyrenees, which contain some very deep and extensive systems including the Reseau de la Pierre Saint-Martin, which for several years was not only the world's deepest known cave (1,342m) but boasted the largest known chamber, La Verna. The system has over 20km of passages and long through-trips are possible. Another fine Pyrenean system is the Reseau de la Coume d'Hyouernedo, which has twenty-eight entrances, over 100km of passage and exceeds 1,000m in depth.

Part of the Salle de la Verna, one of the biggest chambers in the world. (Andy Sparrow)

Germany

There are several major limestone areas and more than 3,000 caves are known. Some systems are well decorated, but few extend over 5km in length. There are a few quite deep caves in the alpine regions of the south, including Geburtstagsschacht at –698m.

Greece

There is much mountainous limestone in Greece and huge potential for discovery. Currently known systems are not long, seldom exceeding 3km, though some are extremely well decorated. In the north, near the Vicos Gorge are the vertical systems of Epos Chasm and the great shaft Provatina, which descends 405m in one huge pitch.

Italy

Italy combines a vast amount of limestone with a long-established tradition of cave exploration. As a result, more than 10,000 caves are known, some over 1,000m deep and others extending in excess of 45km. One popular system, located in the Alps close to the French border, is the Piaggia Bella, which has seven entrances and offers a through-trip over 6km long.

Further south is the Grotta di Monte Cucco, a system over 30km long. The cave consists of a complex of galleries and huge chambers, below which a series of deep shafts descend to –922m. Another great system, and Italy's longest at 53km, is the Complesso Fighiera Farolfi Corchia. Likened in quality to the Gouffre Berger, the three connected caves of Farolfi, Fighiera and Corchia provide a huge and varied system with a total depth of 1,190m.

Scandinavia

There are over 1,500 caves in Sweden, some of which are pseudo-karst systems formed in granite. None is exceptionally deep, but several exceed 2km in length. Many of the systems are formed in marble, as are those in the adjacent country of Norway. Greater depths have been achieved here in systems like the Raggejavreraige, which offers a 620m deep through-trip. The Okshola-Kristihola system has over 11km of passage and a vertical range in excess of 300m. Although few caves in this region have formations, they compensate with fine active streamways cutting through clean and polished rock.

Slovenia

Special mention must be made of this small, newly independent country. This is the home of the classical karst and has outstanding limestone features above and below ground. A special caver's campsite operates from the village of Laze, which is an excellent base. The caves of Planiska and Krizna are fabulous exercises in underground boating and provide unique experiences for cavers at all levels of proficiency. There are two outstanding showcaves: Postojna, which contains a wealth of huge formations, and Skocjanske Jama, one of the largest river passages in the world. There are many deep caves in the alpine areas of the country, including Ceki 2, which descends to –1,502m.

Spain

A good deal of Spain is limestone and there is a profusion of important caving regions across the whole northern fringe of the country. There are a staggering 15,000 known caves and huge potential for further discoveries. There are several systems that descend past 1,000m, including the well-decorated Puerta de Illamina, the very arduous Sima 56, which descends no less than fifty-four pitches, and the classic through-trip of Badalona. These are all in the northern regions, but Sima GESM situated in the south, near Malaga, offers a sporting descent to 1,101m.

Spain is rich in horizontal systems, of which the longest is Ojo Guarena at 110km. Another justly popular system is the Cueto Coventosa, which exceeds 35km in length. One entrance descends a series of pitches to –580m before entering a series of huge chambers and connecting to a long, active river passage and an exit near the resurgence.

There are extensive systems at Matienzo, which is a favourite expedition site for British cavers. Much work has been dedicated here to linking the various caves, which has culminated in the 50km Sistema de los Cuatro Valles.

Off the Spanish mainland, the island of Majorca is composed mainly of limestone and has numerous known caves. Inexpensive off-season holiday packages are making this an increasingly popular venue for visiting cavers.

Switzerland

This is an important caving country and home to one of the world's longest caves – the Holloch. This labyrinthine system exceeds 194km in length and consists mainly of large phreatic tunnels, the lower levels of which are very flood-prone (during one famous incident a group was trapped in the system for ten days). The system runs gradually up-dip and has a total vertical range of 938m.

Another great system is the Siebenhengste, which is 154km long, has thirteen entrances and a depth of 1,340m. There are over 4,000 known caves in Switzerland distributed among extensive limestone regions and much potential exists for major deep systems.

Turkey

There are several karst areas throughout the country but the most important is the Toros Daglari region, where several thousand caves are known. There are some large river sinks and resurgences in this area and water tracing has revealed the longest proven sink to rising link in the world – a staggering 75km. The depth potential has been demonstrated by the exploration of Evren Günay Düdeni, which reaches 1,429m.

Eastern Europe

The Czech Republic has some limestone and one system, Amaterska, exceeds 32km in length. Slovakia has more extensive karst (see above), and the Domica cave crosses under the border with Hungary to become the 25km Baradla Barrang systems, which is well decorated and in parts commercialized. There are more than 1,500 other caves in Hungary, all of which are very varied in nature.

Poland has only a small karst area in the Tatras Mountains, but this has revealed some important caves. The longest and deepest is the difficult, tight and wet Jaskinia Wielka Sniezna at over 22km long and 824m deep.

Romania has some of the best karst areas of Eastern Europe distributed throughout its area. There are several spectacular river caves including the longest currently explored, Pestera Vintului at 42km.

Bulgaria has over 3,000 recorded caves, including systems exceeding 15km in length. Caving is a popular pastime here and there are many clubs in existence throughout the country.

Ukraine

In the western Ukraine is the world's most important known area of gypsum karst and two remarkable caves. The systems of Optymistychna and Ozernaya are incredible concentrated mazes, the latter surveyed to over 225km of passage, making it the third longest known cave in the world.

Georgia

Currently the two deepest caves known in the world are both in Georgia. These are Illyuzia-Mezhonnogo-Snezhnaya at –1,753m and the first, and currently only, cave to exceed 2km in depth, Krubera (Voronja) Cave at –2,191m.

Asia

Beyond the Ural mountains (another caving area of some potential) are the vast limestone regions of central Asia, which extend from the tundra of Siberia to the arid borders of Iran and Afghanistan. Expeditions to the republics of Kazakhstan, Turkmenia and Uzbekistan have revealed immense possibilities, and one system (Boj-Bulok, Uzbekistan) has been bottomed at –1,429m.

The limestone regions of Pakistan and India have failed to reveal any world-class caves, but further east through Thailand and Vietnam discoveries have been spectacular. In Thailand some very large passages have been explored and one system, Tham Nam Khlong Ngu, has what is probably the world's tallest column at 61m. Recent expeditions to Vietnam have explored some huge river passages and surveyed one system to 13km.

China

China has vast areas of limestone (over a million square kilometres) and is one of the world's most exciting caving regions. Much of the surface landscape is tower karst with great clusters of mountainous pinnacles hundreds of metres high. Many huge river caves have been explored, such as the Gebihe system. This begins with an entrance over 100m high into which a large river flows. A series of lakes must be crossed by boat, and at one point a shaft rises 370m to daylight overhead. After 4km a second huge surface shaft is passed before the river sumps. At the resurgence for the system a large entrance leads into the Miao Room, the second largest chamber in the world with a volume of 7–10 million cubic metres. This is just one discovery in what can only be the early days of Chinese cave exploration. The currently longest cave is Shuanghe Dongqun at 117km and it seems certain that this and other systems have many remarkable secrets to reveal.

Middle East

This troubled region has had limited exploration and much work still remains to be done. The Zagros mountains of Iran were investigated during the pre-revolutionary 1970s and a British team descended the deep Ghar Parau until running out of tackle. With high hopes of a depth record they returned the next year to find a sump just beyond their previous limit. The cave had reached

China is a land of caves with many hidden secrets. (Rob Eavis)

barely half of its 1,700m depth potential, and the term to be 'Ghar Paraued' entered the cavers' vocabulary.

Iraq has potential and a few reported caves up to 1km in length, as has Syria and Jordan. Lebanon has much limestone and over 1,000 caves are known. These include the 8km long river cave of Faouar Dara, described as 'unbelievably beautiful', which descends twenty-two pitches among fine formations to reach –622m.

Israel has many small caves including the unusual 3km Malham Cave, which is formed in salt. Yemen and Saudi Arabia have some reported caves but serious investigation has not been possible. Oman has Kahf Hoti, a 5km through cave, and Majlis Al Jinn, one of the biggest chambers in the world, entered by a spectacular 100m daylight pitch.

Borneo

Even greater caves are to be found in the Mulu region of the island of Borneo. Concealed among the dense rainforest are the world's largest known passages and chambers. Gua Payau (Deer Cave) is 100m high and wide and runs for a kilometre, cutting straight through a mountain ridge. Gua Air Jernih (Clearwater Cave) has now exceeded 150km in length and has one of the world's greatest river passages, but the outstanding feature in the region is Sarawak Chamber in the Lubang Nasib Bagus system. This is by far the biggest known chamber in the world and measures 700 by 400m with a height of 100m. It is estimated to have a volume of 12 million cubic metres.

Indonesia and New Guinea

The potential of New Guinea has long been realized and expeditions of several nationalities have made regular visits there. The extensive limestone is combined with exceptionally high rainfall to create some huge river sinks, but the dense jungle and complex terrain make discovery and exploration very difficult.

A British expedition to Irian Jaya in 1989 discovered what may be the biggest river sink in the world, where the Baliem river, with an average flow of 80 cumecs, sinks through boulders. Subsequent exploration revealed a huge system prone to complete and regular flooding by a rise in water level of 150m.

To the east, in Papua New Guinea, exploration has revealed a number of systems including the 54km Mamo Kananda, a complex of streamways and huge fossil galleries. The depth potential has now been partly realized by the discovery of the Muruk system, which descends to –1,141m, the first cave in the southern hemisphere to exceed 1,000m depth.

The island of New Britain has revealed a number of important caves including the classic Nare River Cave. The Nare begins with a huge crater in the jungle giving a 230m pitch into the river cave. The cave sumps after only 2km but the powerful river makes exploration a technical nightmare of traverses and tyroleans.

Australasia

Australia

The major caving areas are concentrated in the south of the continent and include the arid and featureless Nullarbor Plain, which, at 200,000 square kilometres, is one of the world's largest expanses of limestone. The caves of the Nullarbor are entered through dolines and have some very large passages including Old Homestead Cave, described as 'horizontal, easy and dry', which now exceeds 28km. The most famous cave in the area is Cocklebiddy, extending for 6km, most of which is entirely flooded. This is a famous site for long penetration dives, the crystal-clear water creating superb conditions. The longest cave of the continent is the Bullita Cave System at 119km. There are no deep caves on the Australian mainland but the island of Tasmania has the Anne-a-Kananda system, which reaches –373m. The systems of Khazad-Dum and Growling Swallet both exceed 300m in depth, and the latter offers probably the best trip in Australia with a varied 4km traverse between entrances.

New Zealand

There is considerable and varied karst throughout the North and South Islands and local cavers have been very active in exploration. The major systems are Bulmer Cave, which now exceeds 50km, and Nettlebed Cave. Nettlebed was explored from the bottom up to reveal more than 24km of passage before being connected to Blizzard Pot to create a 889m-deep through-trip.

Africa

There are relatively few karst areas throughout the African continent, but the most important, and most accessible to European expeditions, are those in the Atlas Mountains in the extreme north of the continent. Morocco has some fine caves, including the 8km Grotte de Chara, the 17km Wit Tamdoun and the 722m-deep Kef Toghobeit. Further east along the mountain range in Algeria are Africa's longest and deepest systems – Rhar Bouma'za at more than 18km and Anou Ifflis at –1,170m.

Ethiopia has an estimated 100,000 square kilometres of limestone, which has scarcely been examined. The possibilities are graphically illustrated by the cave Sof Omar, which has over 15km of passage including some huge galleries and chambers.

There are small karst areas in several other countries that await further investigation. Gabon has a 1.4km system, Kenya has both limestone and lava tubes, one of the latter exceeding 12km, and areas of Tanzania, Zaire and Zambia have karst regions with recorded caves. In Zimbabwe, deep caves have developed in quartzite and one, Mawenge Mwena, holds the depth record for southern Africa at –305m.

Botswana and Namibia have caves exceeding 1km, but South Africa has much more extensive systems including the well-known showcaves of Cango with their large chambers. The most important karst area lies north-west of Johannesburg and contains the 12km system of Apocalypse Pot, which has a maze of joint controlled passages. There are several other well-decorated and extensive caves in the region.

The greatest potential lies in the island of Madagascar, which has vast, unexplored areas of limestone. Several systems of huge horizontal tunnels have been explored and at least four caves are over 10km in length. The Ambatoharanana system now exceeds 18km and looks likely to become the continent's longest.

North America

Canada

There are a few short systems in the eastern provinces but the most important region is in the Rocky Mountains. Access to this remote area has hampered exploration but several fine systems have been explored. The Nakimu Caves extend for over 5km and have a very spectacular river passage. Arctomys Cave is the deepest in North America at

−536m, most of the descent being by a long inclined streamway. Castleguard Cave extends under the Columbia Icefield and is only explorable during the winter when most surface water freezes. The main passage is over 8km long and offers some very arduous traversing. The entire system has been mapped to more than 18km.

USA

There are twenty major caving regions, and 15 per cent of the total area can be classed as karst. More than 30,000 caves are known, over 350 of which exceed 3km in length, and at least thirty of which exceed 20km. Of the seven longest caves in the world, five are in the USA. That other essential component for

85m waterfall in Mystery Falls Cave, Tennessee. (Peter and Ann Bosted)

the discovery of caves also exists: a large number of active and determined cavers.

The most important karst region extends along the Appalachian Mountains, through West Virginia and into Kentucky. The caves of West Virginia are geologically simple with long, uninterrupted lengths of active and fossil passage. The longest caves in the state are Friar's Hole at 73km and Organ Cave at 63km. There are numerous other long systems and new discoveries are frequently made.

Kentucky is home to the world's longest known cave – the Mammoth Cave System. This currently has 590km of mapped passage, but exploration continues and connections with nearby caves, such as the 177km Fisher Ridge System, are very likely. Mammoth Cave consists of very large fossil

A deep entrance 'pit' – Cagle's Chasm, Tennessee. (Peter and Ann Bosted)

Tom's Lake in California Caverns, California. (Peter and Ann Bosted)

galleries overlying an active system of gently flowing rivers; there are few calcite formations but attractive gypsum deposits are common. The several entrances allow very long through-trips and a five-day, 50km trip has been completed.

Another great system is Jewel Cave in the Black Hills of South Dakota. The cave, named for its fine calcite crystals and gypsum deposits, is an incredible concentrated maze of passages that have been surveyed to a total of 225km. Wind Cave is a similar system in the same region that extends for 208km. The Rocky Mountains have a great number of caves including the second deepest, Columbine Crawl at −429m. There are several other systems over 300m deep, but they lack extensive horizontal passage.

Further south in the Guadalupe Mountains of New Mexico are the two exceptional caves of Lechuguilla and Carlsbad Caverns. These limestone caves have formed by the unusual process of natural sulphuric acid rising under phreatic and artesian pressure. The complex chemistry has resulted in some of the world's most unusual and spectacular cave formations.

Carlsbad Caverns are a nationally famous tourist attraction and at one time claimed the world record for the largest chamber – the superbly decorated 'Big Room'. The system now extends for 44km, but has been overshadowed by its recently discovered neighbour, Lechuguilla Cave. This system, the deepest in the USA at −485m, now extends to 201km and continues to grow. The cave consists of huge galleries and chambers adorned with almost unbelievable formations.

Central America

Mexico

Mexico first attracted cavers from the USA, but now draws expeditions and visiting cavers from around the world. There are vast areas of unexplored limestone and the potential to discover record-breaking caves: one water trace has revealed a potential depth of 2,525m. The longest system is currently the 93km Purificación, which has a rich variety of passages and offers a 895m-deep through-trip.

Sistema Huautla is the deepest known at −1,475m. This is another varied system with 62km of passage and multiple entrances. At the resurgence end of the system, in the Cueva Pena Colorada, divers have revealed over 7km of cave and a series of long sumps.

One of the most famous features of Mexico is the Sótano de las Golandrinas. This huge natural shaft, with a 50m diameter surface opening, bells out dramatically to allow a completely free-hanging descent to the floor 333m below. It is one of the world's classic caving experiences and a place of pilgrimage for visiting cavers.

West Indies and Cuba

Seventy per cent of Cuba is limestone and there are fine examples of karst forms both above and below ground. There are many active caving groups on the island and exploration has revealed a large number of well-decorated caves. The Gran Caverna Santo Tomas has 44km of passage including five long separate river passages.

Andros island in the Bahamas is world renowned amongst cave divers for its 'Blue Holes' – extensive and beautiful caves that are now submerged by the sea. Haiti, Dominica and Puerto Rico have large unexplored karst areas and preliminary investigation has revealed systems exceeding 9km. Caving is more advanced on Jamaica, which is also rich in limestones. Quashies River Cave has been described as an aqueous sporting classic.

The magical Chandelier Ballroom in Lechuguilla Cave. (Peter and Ann Bosted)

South America

The karst of South America compares poorly with that of the northern continent, and that which does exist is often extremely difficult to access. Brazil has more than 1,000 known caves, including the huge and extremely impressive Janelao. Many more systems are presumed to await discovery, deep in the rainforest.

Peru currently has the deepest system in the continent, the Sima de Milpo at –407m, and to the north, in Ecuador, several river caves have been explored. Most other countries have limited limestone and few systems over 1km, but Venezuela is an exception.

Many systems have been explored here, some exceeding 10km. Unusual systems have also developed in sandstone, including the massive surface shaft of Sima Mayor de Sarisarinama, which has 252m pitch.

Oceania

Hawaii

This tropical paradise has the world's four longest lava caves, the greatest of which is the 65km system of Kazumura Cave. Kazumura can also claim to be the USA's deepest cave with a vertical range over 1,100m.

Further Reading

Many cavers, especially those marooned far from their nearest caving area, have limited opportunities to get underground but become enthusiastic readers of caving literature. All aspects of caving from the technical or historical to the humorous have been comprehensively covered by several generations of authors. Most of these books are long out of print but can often be found online. In addition, there are a multitude of caving journals and magazines produced by clubs and societies around the world that contain a wealth of diverse information. The major caving clubs operate journal exchanges and many have their own extensive libraries. There are two principal national caving magazines that contain up-to-date news and various articles of interest. These are *Descent* (Wild Places Publishing), every two months, and *Speleology*, published quarterly by the British Caving Association.

The following list is not comprehensive but provides a selection that covers most topics, such as guidebooks, techniques, pictorial, historical and entertainment.

Guidebooks

Brook, D., and others, *Northern Caves*, new series, 3 vols (Dalesman Publishing, 1988–94). Comprehensive and essential guides to the caves of northern England.

Cooper, M., *Not for the Faint-Hearted* (Purprise Press, 2007). Detailed and practical guide to a selection of fifty harder caving trips in Yorkshire.

Elliot, D., and D. Lawson, *SRT Rigging Guide* (Lizard Speleo-Systems, 1987). SRT rigging in classic Pennine caves. Still very useful but P-anchors have changed some routes.

Gill, D.W., and J.S. Beck, *Caves of the Peak District* (Dalesman Publishing, 1991). Comprehensive and essential guide to the caves and mines of Derbyshire.

Irwin, D.J., and A.R. Jarratt, *Mendip Underground: A Caver's Guide*, 4th edition (Bat Products, 1999). Detailed descriptions of caves of the Mendip Hills of Somerset. Includes stone mines in Wiltshire.

Irwin, D.J., A. Moody and A. Farrant, *Swildons Hole* (Wessex Cave Club, 2007). Exhaustive history and description of this much-loved Mendip cave with excellent photographs.

Marshall, D., and D. Rust, *Selected Caves of Britain and Ireland* (Cordee Ltd, 1997). Good value guidebook to some of Britain's most popular caves.

Middleton, J., and T. Waltham, *The Underground Atlas: A Gazetteer of the World's Cave Regions* (Robert Hale, 1986). Lists the karst regions and major caves in every country of the world.

Mullan, G.J., ed., *Caves of County Clare and South Galway* (University of Bristol

Spelæological Society, 2003). High-quality publication with surveys of major systems.

Stratford, T., *Caves of South Wales* (Cordee Ltd, 1995). Lists and describes the caves of South Wales and the Forest of Dean.

Technique

Elliot, D., *Single Rope Technique: A Training Manual* (Troll Safety Equipment, 1986). Excellent SRT manual for basic rigging and progression.

Farr, M., *Diving in Darkness: Beneath Rock, under Ice, into Wrecks* (Wild Places Publishing, 2003). Cave diving manual by one of the world's most experienced cave divers.

Howes, C., *To Photograph Darkness: The History of Underground and Flash Photography* (Southern Illinois University Press, 1990)

Howes, C., *Images Below: Manual of Underground and Flash Photography* (Wild Places Publishing, 1997)

Judson, D., *Caving Practice and Equipment* (Cordee Ltd, 1995). Some sections are outdated but others remain useful.

Lyon, B., *Venturing Underground: The New Speleo's Guide* (EP Publishing, 1983). Well illustrated and still very relevant with much helpful advice.

Marbach, G., B. Tourte and M. Alspaugh, *Alpine Caving Techniques* (Speleo Projects, 2002). Excellent and exhaustive technical manual covering equipment and techniques.

Meredith, M., and D. Martinez, *Vertical Caving* (Lyon Equipment, 1986). Useful and comprehensive look at SRT techniques.

Warild, *Vertical*, 5th edition (2007), available at http://www.cavediggers.com/vertical/. Comprehensive vertical techniques manual from one of the world's leading cavers.

Historical

Balch, H.E., *The Mendip Caves* (John Wright & Sons, 1947–8). Pioneer exploration under Mendip graphically described and illustrated.

Beck, H.M., *Gaping Gill: 150 Years of Exploration* (Robert Hale, 1984). The story of the exploration of one of Britain's most famous caves.

Beck, H.M., *Beneath the Cloud Forest: A History of Cave Exploration in Papua New Guinea* (Speleo Projects, 2003).

Cadoux, J., *One Thousand Metres Down* (Allen & Unwin, 1957). The story of the discovery and exploration of the Gouffre Berger, the world's first 1,000m-deep cave.

Casteret, N., *Ten Years under the Earth* (Cave Books, 1988). Just one example of the work of the prolific and legendary Casteret.

Chevalier, P., *Subterranean Climbers* (Faber and Faber, 1951). The exploration of the Dent de Crolles system in France. One of the greatest stories of adventure ever told.

Cullingford, C.H.D., *Exploring Caves* (Oxford University Press, 1951). Quaint and readable snapshot of caves and caving in the early days.

Eyre, J., *The Cave Explorers* (Cordee Ltd, 1981). Epic and often hilarious tales of discovery and adventure.

Eyre, J., and J. Frankland, *Race against Time* (Lyon Equipment, 1988). Story of the Cave Rescue Organisation, describing some of the most difficult rescues achieved beneath the Pennines.

Farr, M., *Darkworld: The Secrets of Llangattock Mountain* (Gomer Press, 1997). Fascinating reading with excellent illustrations and photographs.

Farr, M., *The Darkness Beckons: The History*

and *Development of Cave Diving*, 3rd edition (Cave Books, 2000). Extremely well illustrated history of cave diving.

Gemmell, A., and J.O. Myers, *Underground Adventure* (Dalesman Publishing and Blandford Press, 1952). Describes the post-war discovery of some of the classic systems beneath the Pennines.

Heap, D., *Potholing: Beneath the Northern Pennines* (Routledge and Kegan Paul, 1964). Superbly written descriptions of caving back in the days when 'men were men'.

Jasinski, M., and B. Maxwell, *Caves and Caving: A Guide to the Exploration, Geology and Biology of Caves* (Hamlyn, 1967). Techniques may be outdated but has a very good section on cave formation.

Johnson, P., *The History of Mendip Caving* (David and Charles, 1967). Provides a useful overview but seems to have a bias for and against certain clubs.

Rose, D., and R. Gregson, *Beneath the Mountains: Exploring the Deep Caves of the Asturias* (Hodder & Stoughton, 1987). Fascinating and well-told story of exploration beneath the mountains of Spain.

Watson, R., *Under Plowman's Floor* (Zephyrus Press, 1978). A very readable novel about the life of a fictional American caver.

Widmer, U., *Lechuguilla: Jewel of the Underground*, 2nd edition (Speleo Projects, 1998). Fabulous pictorial book of the world's most beautiful cave.

Witcombe, R., *Who was Aveline Anyway?: Mendip's Cave Names Explained*, revised edition (Wessex Cave Club, 2008). If you have ever wondered how caves get their names ....

Useful Websites

Online resources can greatly extend the range of information available. A notable source for facts and links regarding the 'longest and deepest' may be found at www.caverbob.com.

www.trycaving.co.uk	for newcomers to caving
www.caving.uk.com	useful directory of clubs and other services
www.caverescue.org.uk	information on cave rescue
www.metoffice.co.uk	to avoid that flood rescue
www.issa.org.uk	for cave artists
www.ukcaving.com	most popular discussion forum
www.british-caving.org.uk	British Caving Association
www.caveclimb.com	for equipment and training courses

Glossary

abseil   descent by rope usually using a harness and descender.

active cave   cave with flowing water.

adit   horizontal tunnel into a mine.

ammunition box   sturdy waterproof metal case available in various sizes.

anchor   point to which a rope is secured.

aragonite   carbonate mineral often deposited in caves.

ascender   a device that can easily be slid up a tensioned rope and then self-locks in position.

assisted handline   technique for safeguarding and helping cavers on short climbs.

aven   shaft or climb reached from below.

bad air   concentration of noxious gas, usually carbon dioxide.

bang   cavers' term for explosives.

bang fumes   toxic fumes released by explosive.

bat-brake   metal plate that can be used to lifeline.

BCA   British Caving Association.

BCRA   British Cave Research Association.

BCRC   British Cave Rescue Council.

BDH container   waterproof plastic drum manufactured in various sizes.

bedding plane   weakness or crack separating layers of rock.

belay   (n) an anchor point; (v) to lifeline.

belay belt   belt designed to support body weight.

belay plate   small device used by climbers (also known as sticht plate) but not recommended for general caving use.

bend   a knot that connects two ropes.

blue hole   flooded surface shaft leading to underwater caves, especially in shallow marine waters.

bobbin   a simple descender.

bolt   an artificial anchor.

bolting platform   collapsible platform used for scaling avens.

boulder choke   collapsed mass of boulders partially or completely blocking a passage.

boulder ruckle   Mendip term for boulder choke.

bowline   knot used for rope attachment to an anchor.

bowline on the bight   a variation on the bowline that gives two attachment loops.

break-down   collapse of cave roof resulting in boulder-strewn floor.

butterfly knot   a simple knot used for rigging.

c-link   metal clip used for joining lengths of caving ladders.

calcite   most common cave mineral that crystallizes from solution to form stalactites.

Capuchin knot   ball-shaped knot tied into the end of a rope for safety.

carbide   calcium carbide, a chemical that reacts with water to produce acetylene.

carbide lamp   lamp using carbide, which burns acetylene in a naked flame.

carbonic acid   acid formed by the reaction between water and carbon dioxide.

carboniferous geological period from 270–350 million years ago when most British cave bearing limestones were deposited.

cave pearl spherical calcite deposit around grit particle.

CDG Cave Diving Group.

chert natural form of silica often visible as thin black bed in limestone.

chimney (n) a pitch narrow enough to be free-climbed; (v) to free-climb a narrow pitch using two opposing walls.

chockstone stone or rock jammed between two walls.

choke blockage of the cave passage by sediment, stones or boulders.

CIC Cave Instructor's Certificate.

classic abseil method of abseiling without a harness, using the body as a friction brake.

climbing pole a sectional or extending pole used to assist bolting an aven.

clint a surface exposure of horizontal limestone.

clove hitch simple knot that connects a rope to a karabiner.

conglomerate rock composed of broken fragments cemented together by sandy matrix.

cord technique ultra-lightweight system of using one rope lowered and hoisted using cords.

cowstail a short length of rope used as a safety connection.

cumec a water flow of 1 cubic metre of water per second.

curtain sheet-like cave formation developed where water trickles down an overhanging wall.

deads piles of loose rocks left in disused mines.

decorated containing notable displays of speleothems.

descender friction device used for abseiling.

deviation a short cord and karabiner used to redirect a rope.

dig attempt to pass a blockage by excavation.

dip angle of the limestone bed.

doline a funnel-shaped surface depression.

double fisherman's knot secure knot used to join two ropes.

double lifeline lifeline that can be operated from the base of a pitch, enabling every caver to be safeguarded.

duck a passage almost completely filled with water.

dynamic rope rope as used by climbers with substantial stretch and shock absorption.

eco-hanger see *resin anchor*.

electron ladder a lightweight flexible ladder with wire sides and aluminium rungs.

exo-hammock specially designed caving hammock that gives very high insulation.

fall factor the relationship between the length of a fall and the length of the rope – a way of assessing the degree of shock-loading.

false floor a stalagmite floor with a cavity underneath.

fault crack or fissure caused by movement of one rock mass in relation to another.

figure 8 knot knot used for rigging and rope attachment.

flowstone stalactite deposit in the form of a layer or coating of rock surfaces.

fossil remains of living creatures visible within rock.

fossil passage a passage abandoned by the water that formed it.

free climb climb up or down undertaken without the use of ropes.

frog system simple climbing system used for single rope technique.

garda knot self-locking knot using two karabiners.

Gemlock a British-made self-locking descender.

gour   pool formed by a stalagmite deposit around its rim, also know as rimstone pool.

Grigri   a self-locking lifelining device.

grike   fissure formed by enlarged joint in horizontal surface limestone exposure.

gripping   practice of digging ditches on moorland to aid drainage.

gull cave   cave formed by mechanical rather than erosional process.

gypsum   hydrated calcium sulphate, a mineral in which caves can form and which can produce crystal flower formations.

handline   a rope held by hand to assist a short climb.

hanger   metal plate or ring screwed into expansion bolt and used as an anchor.

harness   webbing waist belt and leg loops designed to support body weight in reasonable comfort.

hawserlaid   traditional method of rope construction.

helictite   an eccentric stalactite that grows in any direction.

HMS karabiner   pear-shaped karabiner designed for Italian hitch use.

hydrogen sulphide   toxic gas present in some disused mines with a distinctive rotten eggs smell.

hypothermia   lowering of the body's core temperature by 2 or more degrees, leading to a potentially fatal condition.

Italian hitch   a simple knot tied around a karabiner that can be used for lifelining or abseiling.

jammer   an ascender that grips the rope with a toothed and sprung cam.

joint   crack or fissure that develops at 90 degrees to the bedding and is vertical in horizontal strata.

karabiner   a metal clip with a sprung gate used for connecting ropes and equipment.

karren   solution grooves cut into exposed limestone by running water.

karst   any limestone area showing classic features of limestone land forms.

kernmantle   modern rope construction in which fibres of nylon are contained within a plaited sheath.

klemheist   self-locking knot tied around a rope using a tape or webbing.

ladder   see *electron ladder*.

lapiaz   a chaotic landscape of bare limestone.

lava tube   cave formed in volcanic regions by flowing lava.

LCMLA   Local Cave/Mine Leader Assessment.

leptospirosis   serious illness caused by a bacterium found in some caves.

lifeline   (n) safety line intended to prevent injury by falling; (v) to operate a safety line.

limestone   sedimentary rock composed of at least 50 per cent calcium carbonate.

maillon rapide   a connecting metal link with a screw-sleeved gate manufactured in various shapes and sizes.

mariner's knot   one method of arranging a knot that can be released under load.

master cave   a major passage that conducts the combined drainage of an area.

maypole   a sectional pole used to reach a high-level passage.

meander   a snaking rift passage that may need to be traversed at high level.

molephone   radio telephone system that can transmit through solid rock.

moonmilk   a soft chalky calcium deposit that is usually white.

NCA   National Caving Association.

neoprene   closed-cell foam rubber with good insulating properties.

ogof   Welsh word meaning cave.

overhand knot   a very simple loop knot.

oversuit   a PVC or proofed nylon one-piece overall designed for caving.

oxbow   a loop of passage branching off and then rejoining the main route.

paragenesis   process of upward erosion in a phreatic passage where sediment deposits prevent erosion of the floor.

phreas   a permanently flooded zone.

phreatic   cave or passage formed in the phreas, i.e. while entirely flooded.

pitch   a vertical drop requiring tackle.

polje   an alluvial floored basin surrounded by limestone hills.

ponor   sinkhole taking water from a polje.

pontoniere   a chest-high latex garment that keeps the wearer dry when wading in deep water.

pot   pothole.

pothole   literally a deep hole etched in a river bed by rotating cobbles, but often used to describe a vertical shaft or cave entered by, or containing, one or more shafts.

potholing   strictly, the exploration of vertical caves but generally used to denote any cave exploration.

prusik   to ascend a single rope using special knots or ascenders.

pseudo-karst   a non-limestone area that shows limestone characteristics, especially cave formation.

pulley jammer   a jammer and pulley linked together.

pull-through   technique of abseiling down a pitch and retrieving the rope from below.

radon   a naturally occurring radioactive gas that accumulates in some caves.

rain shadow   effect whereby hillside facing into wind receives much greater rainfall and vice versa.

rawlbolt   sturdy metal expansion bolt sometimes used as anchor point.

rebelay   point where rope is re-anchored midpitch.

resin anchor   stainless steel anchor held in a drilled hole by a resin, also known as an eco-hanger.

resurgence   point where cave waters emerge on to the surface, also known as rising.

rig   to arrange ropes and equipment enabling a traverse, descent or ascent.

rimstone pool   see *gour*.

rising   see *resurgence*.

rope protector   a pad used to protect a rope from abrasion with the rock.

rope walking   a very fast method of ascending a single rope.

ruckle   Somerset term for a boulder choke.

scallops   regular hollows in a cave wall caused by water flow.

self lifeline   protect oneself while climbing by connecting to a rope with an ascender.

Shetland attack pony   electronic compass and clinometers used in cave surveying.

shunt   an ascender that can be used to safeguard abseil descents.

single rope controller   a simple device used for lifelining.

single rope technique   system of exploring caves with single ropes on pitches using abseiling and prusiking.

sink hole   point where water sinks, or may have sunk.

slave   remote-control unit for firing a flashgun.

sling   a loop of rope or webbing.

speleology   the scientific study of caves.

speleothem   any cave formation, i.e. stalactite, curtain, gour pool, and so on.

spit   a small expansion anchor that can be installed by hand using a special tool and a hammer.

squeeze   a constriction in the passage making it only just passable.

SRT   single rope techniques.

stalactite   a cave formation projecting downwards from the roof.

stalagmite   a cave formation growing upwards from the floor.

static rope   a low-stretch rope used for general-purpose caving and especially for SRT.

sticht plate   see *belay plate*.

stop descender   auto-locking descender manufactured by the French company Petzl.

stope   large vertical fissure in a mine often made dangerous by rotting false floors.

straw   a tubular stalactite equal in diameter to a single drip of water.

strike   direction along a bedding plane at right angles to the angle of dip.

sump   a completely flooded passage.

survey   a map or diagrammatic representation of a cave.

survival bag   lightweight plastic bag used to retain body heat in an emergency.

swallet   South of England term for a sink hole.

syphon   a sump.

tackle   ropes, ladders and any other rigging equipment.

tackle bag   bag designed to transport tackle.

tape   very strong webbing intended for caving and climbing use.

through-trip   trip where entry and exit are from different entrances.

topofil   surveying instrument used by many overseas cavers.

trap   term occasionally used for a sump.

traverse   a passage that must be followed above floor level; to climb horizontally.

troglobite   animals that live permanently underground and have adapted to the environment.

troglophile   animals that live underground through choice and have no special adaptation.

trogloxene   animals that are occasional or temporary cave dwellers.

undersuit   fleece or fibre pile one-piece suit.

vadose   cave or passage formed by water flowing under normal atmospheric pressure.

water-table   level below which all cavities are flooded – the phreas.

Weil's disease   most virulent and dangerous form of leptospirosis infection.

wet socks   neoprene socks.

wetsuit   neoprene suit worn for warmth in very wet caves.

whaletail   descender commonly used in Australia.

Y hang   a rope arrangement sharing the load evenly between two anchors.

Z-rig   a simply arranged 3:1 hauling system.

Index